# UNDER THE HOOD

# OF THE UNIVERSE

## COOROO EGAN

# UNDER THE HOOD OF THE UNIVERSE

## CONTENTS

INTRODUCTION                                             1


1. LOW TEMPERATURES

    Negative Kelvin                          3

    Bose-Einstein condensates                5

    Superconductors                          7

2. THE BAND GAP

    Spectroscopy                             9

    Solar power                             10

    Lasers                                  12

3. THE QUANTUM WORLD

    Superposition                           14

    Bell's Theorem                          17

    Entanglement                            19

    Quantum computing                       20

**4. TIME**

           Light speed            22

           Special relativity            23

           General relativity            26

           The arrow of time            27

**5. SPACE**

           Vacuum energy            29

           Black holes            30

           Quasars            32

           The Higgs field            33

**6. MISCONCEPTIONS**

           Tachyons            36

           Plutonium            37

           The big bang            38

           Quantum tunnelling            39

**7. ENERGY**

           Nuclear fusion            42

           Nuclear fission            45

           Conservation of energy            46

           Negative energy            47

           $E=mc^2$            48

## 8. THE UNIVERSE

| | |
|---|---|
| Dark matter/energy | 50 |
| Size of the universe | 51 |
| Energy of the universe | 54 |
| Supervoids | 55 |
| Eras | 57 |
| Expanding space | 59 |

## 9. SPIN

| | |
|---|---|
| Tennis racket theorem | 60 |
| The Magnus effect | 61 |
| Rattlebacks | 62 |
| Discs | 63 |
| Particle spin | 64 |

## 10. EVERYDAY THINGS

| | |
|---|---|
| Pendulums | 67 |
| Surface tension | 68 |
| Ice skating | 70 |
| Fire | 71 |
| Water | 73 |

## 11. EARTH'S OCEANS

The Great Pacific garbage patch     76

Waterfalls     77

Brinicles     78

## 12. ASTRONOMY

Doppler shift     80

Supernovae     81

Exoplanets     83

Starshades     85

Gravitational waves     86

## 13. EARTH'S MOON

Synestia     90

Tidal locking     91

Earth's 'second moon'     92

Helium-3     93

## 14. RADIOACTIVITY

Carbon dating     95

Radiation     96

Radiation damage     97

The linear hypothesis     98

Backscatter X-rays     100

## 15. PROTECTION

Radiation tablets                    102

Body armour                          103

Faraday cages                        105

## 16. CATEGORIES

Nuclear weapons                      107

Star types                           108

Pluto                                111

## 17. WAVES

Sonic booms                          113

Cherenkov radiation                  114

Sonoluminescence                     116

Sound channels                       118

## 18. NATURAL DISASTERS

Hypernovae                           120

Volcanoes                            121

Tsunamis                             123

Asteroids                            125

Earthquakes                          127

## 19. MATERIALS

Night-sky cooling 130

Strength 131

Graphene 133

Aerogels 134

Metallic hydrogen 135

## 20. MATTER

Plasma 137

Liquids 138

Anti-matter 140

## 21. LIGHT

Mirages 143

Reflection 144

Holograms 145

A random walk 146

Pink 147

## 22. PARTICLE PHYSICS

| | |
|---|---|
| String theory | 149 |
| The LHC | 151 |
| Quantum chromodynamics | 153 |
| Quantum electrodynamics | 157 |
| Neutrinos | 159 |
| Pentaquarks | 160 |

## 23. GRAVITY

| | |
|---|---|
| Rockets | 162 |
| Unifying forces | 163 |
| Space-time curvature | 165 |
| The graviton | 166 |

## 24. MAGNETISM

| | |
|---|---|
| MHD propulsion | 168 |
| Ferrofluids | 169 |
| Permanent magnets | 170 |
| Destroying a magnet | 172 |
| Electromagnets | 172 |
| Lenz's law | 174 |

## 25. EARTH'S ATMOSPHERE

| | |
|---|---|
| Solar flares | 176 |
| Atmospheric phenomena | 177 |
| Shadow bands | 178 |
| Rainbows | 179 |
| Radiometric dating | 181 |
| Global warming | 182 |

## 26. NUMBER CRUNCHING

| | |
|---|---|
| Entropy | 184 |
| Natural units | 185 |
| Big K | 186 |
| The parsec | 188 |
| Calories | 189 |

## 27. A TOUCH OF PHILOSOPHY

| | |
|---|---|
| Planck units | 191 |
| The anthropic principle | 193 |
| Ridiculous extremes | 194 |
| Tesseracts | 197 |

**28. THOUGHT EXPERIMENTS IN SPACE**

               Kugelblitzes                199

               The Drake equation        200

               The Oort cloud         202

**29. DABBLING IN SPACE**

               Voyager                  204

               Cold welding          206

               The smell of space     208

**30. END ON A JOKE**        210

**ACKNOWLEDGEMENTS**     211

# Introduction

When a child picks up a crayon and starts to draw, some parents may envisage a future artist or architect. If they run and jump and climb on every surface, perhaps the parents might see an athlete in the making. But what of the child who takes a toy train in each hand and continues to smash them into each other, just to see what happens? Give them a new train, and seconds later they're smashing that one too. In much the same way, these children may just be future particle physicists.

A particle physicist's job is to try and make sense of the universe on the smallest of scales. They use accelerators instead of hands, smashing particles together instead of toy trains, but no matter how many times they do it, after searching through the debris, you can bet it won't be long until some very similar logic to that of the child is applied and the question is asked... 'What if we were to cause an even bigger smash next time?'

Physics is not confined to the scale of the very small. Being the science that attempts to look under the hood of the universe and see how it ticks, it also deals with concepts on the grandest of scales and everything in between. At first glance this can seem overwhelming. Combine this with a reputation for requiring a lot of mathematics, and many people never scrape the surface. This book is an attempt to rectify that. Consider it a guided whirlwind tour through a selection from the world of physics, hopefully allowing anyone interested a look at the often quirky, sometimes funny, and regularly bizarre sides that the subject has to offer.

# 1. Low Temperatures

**Negative Kelvin**

Two strange things can happen when talking physics. The first is that it's perfectly common to put aside everything that has been previously taught in order to continue. For example, in primary school you may have learned that light travels in straight lines. In high school you may have continued to discover that actually light does bend. If you continued your studies further, you may have been taught that actually light does travel in straight lines, space bends, and so on. That isn't to say that what was being taught earlier was wrong, it was useful for grasping the concepts, just an incomplete picture. The second strange thing is that loopholes in hard and fast rules are perfectly acceptable. For example, it's a well-known rule that nothing can travel faster than the speed of light. Technically the rule is that nothing can travel 'through space' faster than the speed of light. The loophole? Space can expand faster than the speed of light. This can actually be argued well, for or against, once units of measurement are introduced, but it's just an example, and no; technology isn't at a point where we can expand/contract space at will, so we won't be using this loophole to build any star-trek style warp drive systems any time soon sorry, but there are many other loopholes out there that do get used. A great example of an incomplete picture and loopholes coming together can be seen with a quick look at temperature.

Most people are familiar either with Celsius (C) or Fahrenheit (F), possibly having a picture in their head of what temperature means. A higher temperature corresponds to more energy, so if you heat up water, its atoms can gain enough energy that you get steam. Likewise a lower temperature corresponds to less energy, so if you cool down water, its atoms can lose so much energy that you get ice. In physics, temperature is measured in Kelvin (K). The scale starts at a point known as absolute zero, a theoretical point so cold that all motion would stop, and it goes up from there. Our hard and fast rule here is that nothing can be at absolute zero, no matter if it's in a lab or the vacuum of space, you can get close, but nothing can ever actually be at zero Kelvin. So what's the loophole? Well, the rule doesn't prohibit something from being a negative temperature. This may sound crazy, but it's been done successfully, so clearly it's time to put aside the old picture of temperature and introduce a more complete one.

If you were to ask for the cut and dried version of the explanation, it would involve the concept of entropy. Entropy is a measure of the order or disorder of a system. More entropy simply means more configurations, so for example steam has many possible configurations of its atoms and hence a high entropy, while ice, whose atoms are considerably more uniform has a lower number of possible atom configurations and therefore a lower entropy. For any substance at a positive temperature, if you add energy, the entropy increases. For a substance with a negative temperature, the opposite is true, so if you add energy, you decrease entropy. And while that explanation would be technically correct, it probably wouldn't really give a feel for what's happening, so as an alternative, imagine that every atom in a substance has a varied selection of different energy states, perhaps represented as hooks hanging from rungs on a ladder. If you heat up the substance, some of the hooks move up to higher rungs. If you cool it down, some of the hooks move down. The crucial concept here is the distribution of the hooks. No matter how much you heat or cool the substance, most of the hooks are always going to be found on rungs that are lower down the ladder with only a very few scattered further up. However, with a substance at a negative temperature, this distribution is reversed, so no matter how much you heat or cool the substance, you would find the bulk of the hooks hanging from rungs that are higher up the ladder, with only a very few scattered down low. This does all lead to one interesting dynamic. Energy will 'always' flow from a substance at negative temperatures to one at positive temperatures. If talking in the classical sense of temperature again, you could say that a substance at negative temperatures will always be 'hotter' than any substance at positive temperatures, regardless of how hot that might be.

## Bose-Einstein condensates

The familiars of solid, liquid and gas aren't the only possibilities for substances, there are more. Continuing past gas you encounter plasma, but possibly more interestingly, continuing in the other direction past solid you can encounter something called a Bose-Einstein condensate (named after Satyendra Nath Bose and Albert Einstein). To understand why this state is so interesting requires some quick info from quantum mechanics.

If the operating system of the universe was windows desktop, then quantum mechanics would be computer code. An underlying system which governs everything, but tends to stay out of sight. One of the hard and fast rules within quantum mechanics is that the more you know about something's position, the less you know about its momentum and vice-versa. That is, it's absolutely impossible to completely know everything about where something is as well as everything about how fast it's going. The more accurately you measure one, the less you know about the other. This is part of a larger set of rules called Heisenberg's uncertainty principle.

Keeping this correlation between position and momentum in mind, what happens if you cool something right down to a fraction of a fraction above absolute zero? Well, we now know the momentum of the atoms involved extremely well, they're a fraction of a fraction off not moving at all. So what does this mean about the position of these atoms? It must mean that they could be anywhere within a certain area. Don't confuse this with them simply playing hide and seek in an area waiting to be found, their position according to the universe is actually uncertain. And to fully highlight this, what happens if we keep cramming atoms with uncertain positions into the same space? Well it turns out that you can actually put more than can fit into the one area, their possible positions can overlap, but that's fine. This counter-intuitive substance then is what we call a Bose-Einstein condensate, and was the first substance to be used to slow down and stop light in its tracks before sending it on its way again.

## Superconductors

Conductors are a part of modern everyday life, whether it's the filament inside your light bulb, or the power lines leading to your house. They conduct electricity by allowing the flow of electrons from one end to the other. Add an electron on one end, and the electrons all shuffle along so that one falls off at the far end. Different materials are selected for different applications, so in a light bulb you want a certain amount of resistance so the electrons lose a portion of their energy to the atoms they encounter along the way, causing the filament to heat up and produce light. In power lines, you want to minimise this resistance in order to maximise the energy reaching the final destination. When the temperature is lowered enough, some materials are so efficient at this that they have zero resistance. They have absolutely zero loss of energy from the electrons to the atoms they're jumping between, allowing the electrons to continue indefinitely. These materials are known as superconductors.

Superconductors exploit something known as the band gap, also known as the energy gap. The electrons orbiting an atom can only ever have certain energies. Different atoms allow different energy levels, but for any particular atom, only electrons of set energy values can ever be in orbit. If an electron changes energy levels, it performs what's known as a quantum leap. If it was losing energy, it emits the difference as a photon of light, jumping this energy gap to a lower allowed energy level (an orbit closer to the atom's nucleus). In a superconductor, the atoms are at a very low energy level called the ground state. At this energy, if they would collide

7

with something, say the next atom, the energy from the collision would not be enough to perform the quantum leap to the next allowed energy level, it would put them in the gap between allowed energy levels, the energy gap. As electrons cannot be in this gap, the amazing result is that they don't collide with anything at all. They encounter zero resistance, allowing them to continue moving indefinitely.

# 2. The band gap

## Spectroscopy

So electrons can't be in the band gap, but why? Why is it that electrons orbiting an atom's nucleus are only allowed certain energy levels? Electrons are particle-waves, and it's their wave-type properties that determine what energy levels they're allowed to have within an atom. Imagine the electron's path as a wavy line going around an atom's nucleus, steadily going up and down as it goes around, kind of like a horse on a carousel. If after one lap of the nucleus such a traced path returns it to exactly where it started, then you have a complete wave function and all is well. However, if after one lap the traced path is not exactly where it started, being either higher or lower, then the wave function wouldn't be complete. This can't happen. If it helps to understand why, imagine a hypothetical electron travelling this incomplete path. It would continue around the atom, and it's continued wave function would cause interference, but it would be interfering with itself, resulting in it cancelling itself out of existence. This doesn't happen. No incomplete waves allowed in quantum mechanics. Electrons can never be in an orbit which would cause them to interfere with themselves, and as such, there are a very limited number of acceptable orbits that electrons can ever be in. Another extremely useful application of this fact is spectroscopy.

It was once claimed that humans would never know what the stars were made of. This prediction was possibly made more famous than it otherwise would have been due to its unfortunate timing. Very shortly after it had been made, starlight was put through a prism, the resulting rainbow was analysed and the answer was known, hydrogen. This was exploiting the fact that every atom has subtly differing and unique allowed electron orbits. The upshot being that different elements emit and absorb subtly differing and unique wavelengths of light. Compare the spectral lines of the light after splitting it in a prism, a substance's spectral fingerprint, against those of known substances, and when you see a match, you've determined exactly what something is made of. Useful for everything from analysing the Sun and stars, to smoke stacks of factories you suspect are breaking regulations and everything in between.

## Solar power

The Sun provides the Earth with huge amounts of energy, the trick is being able to convert it into something useful. Plants have been photosynthesizing sunlight into chemical energy for a little under half a billion years now, but even they were far from first on the scene with this ability. Algae had already been photosynthesizing for over half a billion years when the first land plants came along, and there's some evidence that bacteria were photosynthesizing for over two billion years before algae. Humans then are relative latecomers to the idea of energy from sunlight, but in recent decades, we've certainly been making up for lost time. In 1987, there was the first 'world solar challenge', a now biennial solar powered car race across the middle of Australia with the aim of building a solar powered vehicle capable of decent and sustained speeds. By 2005, not even two decades later, many entrants were having to slow down for 110 km/h (~68 mph) speed limits encountered along the way. The race keeps introducing restrictions on things like batteries and solar panels to address this, but solar cells, or photovoltaic (PV) cells, continue to increase in efficiency faster than many would have thought possible, but how do they actually work?

Inside a PV cell, there are two different types of very slightly impure silicon. N-type (for negative), where the impurities cause a slight excess of electrons, and p-type (for positive), where the impurities cause a slight lack of electrons. When brought together, extra electrons from the n-type silicon go over to the p-type silicon. This leads to a charge imbalance at the border creating an electric field. This charge imbalance would normally dissipate except for the fact that silicon is a semi-conductor. This means that it can act as either an insulator or a conductor depending on which energy level its electrons are at. In the PV cell, the electrons are in the valence band, a low energy state where the silicon holds tightly to its electrons and acts as an insulator, thereby allowing it to maintain this charge imbalance. In this configuration, when a photon of sunlight with sufficient energy hits the silicon and provides an electron with the energy needed to perform a quantum leap, jumping the band gap to a higher energy level, the conduction band, where the silicon doesn't hold on so tightly to its electrons and instead acts as a conductor, the electric charge funnels the excited electron in one direction, creating an electric current.

While solar power has many advantages over other power options a lot of the time, there is at least one situation where it has its own unique disadvantage. The Mars rover 'Opportunity' has to wait for gusts of wind to clean the Martian dust off its solar panels before it can get the sunlight it needs to continue on mission, meaning that while it's already lasted much longer than originally envisaged, the chances for its operation have been quite sporadic. Generally speaking though, drawbacks to solar power are largely non-existent, encouraging its continued and fast paced technological development, though there's no real telling which direction that will take in the future. Perhaps the solar farms in Spain which merely focus the Sun's rays with mirrors onto a small area to run a generator will dominate power generation. Perhaps the promise of cheap printable solar cells will eventuate, transforming solar cell use in households from primarily rooftop installations to things like camping tents. There's even been the development of transparent solar cells, meaning windows could double as energy collectors. Whichever direction the technology ends up going though, it looks like Leonardo Da Vinci's 1447 prediction of a future 'solar industrialization' is coming true.


## Lasers

Probably one of the most useful inventions in physics of recent times is the laser (Light Amplification by Stimulated Emission of

Radiation). It works by taking a crystal or gas and giving energy to its electrons, exciting them into a higher energy level. When one of these electrons loses energy and falls back through this energy gap to its original energy level, which will happen naturally if given time, it emits a photon of light. The trick is that thanks to a quantum mechanical effect called stimulated emission, an emitted photon will trigger another electron to also lose its energy and emit a photon, but of the exact same frequency and direction as the first. These two photons then trigger two more electrons to emit photons, again with identical frequency and direction. By the time the original photon reaches the end of the laser, this cascade effect has caused a huge number of electrons to lose their extra energy and emit photons of identical frequency and direction. Physicists say these photons are coherent. The fact that the photons are of identical direction, or are spatially coherent, allows a huge range of applications, from laser pointers illuminating spots over extremely long distances with little loss of brightness, to the focusing of extremely large amounts of energy on very small points for laser cutting. Counter-intuitively, lasers can even be used when getting substances down to extremely cold temperatures by shooting off the more excited electrons, reducing the overall energy (and therefore temperature) of the substance in question.

# 3. The quantum world

## Superposition

It may be easy to dismiss quantum oddities as quirks confined to a world which never interacts with the one we're familiar with. After all, surely in the macroscopic, visible world, seemingly crazy effects must cease, allowing common sense to dominate once more right? Well, not entirely, and possibly no experiment along these lines is more elegant than the dual slit experiment.

Take a thin piece of cardboard or similar, cut two close-together slits into it, put a sheet of paper a little behind at the back, aim a laser pointer at the slits from the front and look at the pattern that the light makes on the paper after passing through the two slits. With two vertical slits, you should see a horizontal stretch of light with bars, portions where there is light, and portions where there's no light at all. What you're seeing is an interference pattern. We can surmise that light is acting exactly as a wave, being diffracted, with peaks and troughs, high points and low points. If the light waves from both slits are peaking at the same time when they reach a point on the paper, you see light. If the light wave from one slit is at a peak and the light wave from the other is in a trough, they cancel each other out, and you see no light on the paper at the back. The same experiment can be done in a pool of water with a wave generator.

Let's now imagine a slight change to this experiment. Instead of firing a beam of light at the slits, we'll have a machine that fires a beam of electrons, with some paper at the back that lights up whenever an electron hits it. Easy enough to do with the right equipment, and once again, as you might expect, we get the exact same interference pattern. But, what happens if we slow down our electron firing machine? We'll slow it down so much that it's firing individual electrons at the two slits, then we'll mark down on the paper wherever an electron hits and repeat until we have a large sample size. You may be surprised to discover that the pattern we get at the back is still an interference pattern. That is to say that a single electron seems to be interfering with itself. This may seem a little strange, but before any explanation, the experiment has one final twist to offer. What happens if we watch the slits really carefully? We'll do the exact same experiment, but install some kind of monitoring device at both slits so that we can tell exactly which slit each electron passes through. We'll again mark down where it hits on the paper at the back, collect a large sample size and see what we get. It turns out if we do this, that the interference pattern completely disappears. The simple act of watching for which slit the electron passes through drastically changes the outcome of the experiment.

So what's happening? Is the electron a particle or is it a wave? The answer is neither, it's a particle-wave. There are a couple of fancy names and terms that can be thrown around here. You may hear things like 'wave-particle duality' (all this is saying is that the light and the electron can seem to change between acting as a particle or acting as a wave depending on circumstance, essentially just sticking a name to the phenomenon). Some people may like to talk about 'collapsing the wave function' (again, just another way of naming what's happening). To understand it though requires a coming to grips with a fundamental fact from quantum mechanics, that the location of a subatomic particle is not defined until it is observed. It has a dual potential nature (wave or particle), but its observed nature must be one or the other. Physicists call this a superposition of states.

The idea of superposition may seem complex at first glance, but can possibly be made simpler when put into everyday terms. Imagine you're in a queue for a ride, perhaps on some bumper cars. When someone reaches the front of the queue, an attendant ushers the person into one of ten cars. Now, before you reach the front of the queue, it could be argued that you're in a superposition of states. You are definitely going to end up in one of the ten ride cars, but you have no idea which until your time comes at the front of the queue. Likewise, our electron is definitely either going to act like a wave, passing through both slits to produce an interference pattern with itself, or it's going to act like a particle, passing through just one of the two slits, and not produce an interference pattern; but which of these options isn't decided until the time comes. Ok, so there are ways you could possibly affect which car you'll end up in from the queue, but it's just an analogy, you're not a quantum particle, if you were, that wouldn't be an option. In the quantum world, whether talking about photons of light, electrons, or some other sub-atomic particle, the default setting is this state of superposition where both particle or wave-like behaviour is possible, but only one can be observed.

# Bell's Theorem

So you may be thinking that the dual slit experiment sounded interesting, but wondering if perhaps there's an experiment you can more easily do at home that will still allow you to see quantum mechanical effects manifesting? Well you're in luck. All you'll need is three pairs of sunglasses (less if you're happy to break them).

Sunglasses have lenses which are polarizing filters. It can be easy to think of these as only letting light waves through which match the direction of the polarization, so the filter may be blocking all light waves of an up/down polarization, and letting through all ones of a left/right polarization, but as we'll see, this isn't exactly the case. So first, take one lens and hold it up to the light. As you'd expect, some of the light will be blocked. If you take another lens and hold it in front of the first one, by rotating it you should notice that varying amounts of light gets blocked. When the two polarizing filters are at 90° to each other, no light should be getting through. Still easy to understand with our initial assumption, as now, whether our light is polarized up/down, or left/right, it's getting blocked by one of the filters. With all possibilities blocked, no light can get through right? Well now for the crux of the experiment. Leave the first two filters in this position, where no light gets through, and introduce a third filter between the two, such that the new filter is at a 45° angle to both the other two filters. You should immediately notice something rather strange occurring. Light is now coming through where no light was before. We've added an extra filter, and yet this is somehow allowing light to pass through where none was before, so what's happening?

The answer has to do with probabilities. The polarizing filter of sunglasses is probabilistic in nature, letting either all or none of a photon through some given percentage of the time. With two filters orientated 90° to each other, the second filter lets through 0 percent of the photons that passed through the first. With another filter in between though, a percentage, but not all of the photons passing through the first filter are stopped at the second, and then a percentage of these are stopped by the third filter, neither percentage is zero though, so not all the light is blocked. Hopefully that made sense, because much like the end of the dual slit experiment, this experiment now has one final quantum mechanical twist to offer.

You might expect that the numbers are consistent. After all, a filter at 90° to the first lets 0 percent of photons through, while at 45° it lets 50% of photons through. So what would a filter at 22.5° to the first do? While you may have been hoping the answer would be a nice predictable 75%, the answer is in fact 85%. Only 15% of photons are blocked. That's not a mistake, the numbers quickly start to not add up the way one might like. This experiment highlights something known as Bell's theorem, which basically states that in quantum mechanics, there are no local hidden variables. In this case, an incoming photon can't have some hidden variable, which, if known, would allow us to determine with certainty its state with respect to each filter before it reaches them. It's not the case that there might be such a variable and it's unknown or even somehow unknowable to us. Such a variable simply cannot exist.

## Entanglement

In everyday life, it doesn't matter what order you ask questions in, you'll get the same answer. Where's the pendulum? How fast is it going? It's pointing down, it's stationary. How fast is the pendulum going? Where is it? It's stationary, it's pointing down. Logical and consistent. But as you may be getting a feel for now, in the quantum world, it matters what order you ask. Remember the uncertainty principle. If you measure where something is, you lose your ability to measure how fast it's going. Physicists say that the operations don't commute. That is to say that taking a measurement of a quantum system must fundamentally affect that system. This leads us to an interesting possibility. First, entangle two quantum objects. With the right equipment, this isn't too difficult. Perhaps you have a device that splits a photon of light into two, where both resulting photons must have the same polarization (either vertical or horizontal). You haven't checked, so until you do, they're in a state of superposition and could be either, but if you measure one as vertically polarized, then the other must also be vertically polarized, and likewise if you measure one as horizontally polarized, then the other must also be horizontally polarized. Now you can separate these two photons of light with any distance, and they will remain fundamentally linked like this. They could be in separate rooms, or separate galaxies, and still, if you take a measurement of one to determine its polarization, it's no longer in a state of superposition, which means that its entangled partner is also no longer in a state of superposition. You've determined something about the other's polarization regardless of the distance involved. It's the instantaneous nature of this phenomenon which led Einstein to refer to it as 'spooky'.

## Quantum Computing

Ever noticed how quickly computers can become outdated? In the span of one lifetime, computers went from figuring out tic-tac-toe, to beating grandmasters in chess, then *Go*, the most complex strategy game we have, to now being able to play and beat humans at real time strategy games like starcraft.

In the latter half of the 20th century, it was observed that chip performance seemed to be doubling roughly every 2 years. This became known as Moore's law and has acted as a kind of self-fulfilling prophecy until today as industries set future targets based off the predictions this made. No, this isn't some law of nature or rule of the universe, so what's it got to do with physics? Well looking ahead to the future, there are certain limitations looming. For instance once you start talking about transistors on the atomic level you literally start running out of room, you physically can't fit more into the space given, but there is an approach being researched which could keep the trend of Moore's law going steadily for a while to come yet, quantum computing.

The idea is simple enough, computers store data in bits. A normal Computer's bits can be either a 1 or a 0, a quantum computer utilizes some of the quantum mechanical quirks already covered to have bits which can be in a superposition of states. These quantum bits, or qubits, can then be a 1 'and' a 0 at the same time, in theory allowing calculations to be done which normal computers just couldn't handle. There are a variety of problems to overcome with the development of quantum computers which normal computers never had to deal with, for instance if the system isn't properly isolated from the outside world, the qubits can lose their state of superposition, called quantum decoherence, but progress in solving such issues is continually being made, and while it may still be some time before you start to see quantum computers in the home, this is one developing technology which it's probably safe to expect some very impressive things from down the track.

# 4. Time

## Lightspeed

One of the most well-known rules in all of physics is that nothing can travel through space faster than the speed of light, and alongside this rule it is often mentioned that time slows down as you approach the speed of light. You may feel you have a handle on that, or you may feel completely in the dark with such things, but either way, the connection is possibly a lot simpler than you might guess. What if the claim were that in fact 'everything' is travelling at the speed of light? You, this book, the car zooming past outside, the photon of light from the lamp next to you, all moving at the same speed. It probably sounds crazy on the surface. Why, you might then ask, if you and a racecar have a race, do you lose every time? Why can't you catch up to the racecar speeding away from you if you're going at the same speed?

The answer is that more of your speed is being subverted into travelling through time. Imagine a very simple graph, with the horizontal x axis being space, and the vertical y axis being time. Now add a single arrow over the top of this graph, pinned to where the x and y axis meet. The arrow can swivel, but is always the same arrow and as such is always the same length. For a photon of light, this arrow is lying horizontal, flat against the x axis, 100% of its speed going into travelling through space. Your arrow is resting almost, but not quite, vertical, quite close to the Y axis, so most of your speed is going into travelling through time (otherwise known as aging). The arrow is the same length for everyone and everything, just slightly swivelled. So as something speeds up towards lightspeed, the arrow slowly swivels down towards the X axis, so less of its speed gets subverted into travelling through time (time slows down). As something slows down towards stationary, the arrow slowly swivels up towards the Y axis, and more of its speed goes into travelling through time (time speeds up). In day to day life, this difference is so slight as to generally be imperceptible, but yes, this does mean that a photon of light never ages. It's exactly the same age when it reaches your eye as the moment when it was first created.


## Special relativity

The speed of light through a vacuum is a constant regardless of the observer's speed. This means that no matter how fast you're going, a beam of light travelling away from you will always seem to be travelling at light speed away from you; you can never even start to catch it, no matter how much you speed up. And that's not all. While time slows down as something speeds up towards the speed of light, two other interesting things also happen, mass increases, and the object becomes squished in the direction of travel, known as Lorentz contraction, so for example a sphere would become more pancake shaped. What can be tricky to get your head around is that things are relative (hence the name, the theory of relativity) to the observer's frame of reference.

To highlight what all this actually means, we'll look at just two classic examples. The first is of a train going through a tunnel. Let's say the train if stopped is the same size as the tunnel. If the train is going 99% the speed of light, then from the perspective of the tunnel, the train has become squished, Lorentz contracted to around $1/7^{th}$ of its size, small enough to now completely fit inside the tunnel. When the train then enters the tunnel, we simultaneously slam down a door on both ends, containing the train briefly within the tunnel (before it smashes through the front door and continues on its way). All may seem fine, but now let's look at it from the perspective of the train. The train doesn't see itself as having contracted, in fact relative to it, everything else, including the tunnel has contracted, so to it the tunnel is now only $1/7^{th}$ the size it was before, tiny compared to the full-size train, and far too small to fit inside. So what happens from the train's perspective? The answer lies in the idea of simultaneity. What was simultaneous from the tunnel's perspective (the slamming closed of the two tunnel entrances) is not simultaneous from the train's perspective. For it, the front end of the tunnel slams closed first, the train crashes through, and then afterwards, once the entirety of the train passes the back end of the tunnel, the second door slams closed behind it. Notice that all the important facts between both versions of events still agree. The front of the tunnel closed causing the train to crash through in both frames of reference. The back of the tunnel closed just after the train had passed that point. Was the train ever completely inside the tunnel? From the tunnel's perspective it was; from the train's, it wasn't. Both perspectives are completely correct.

The second example is known as the twin paradox (though the word paradox in the name is a little misleading). Imagine you have two twins the same age, twin 1 stays on Earth, while twin 2 shoots off into space on a rocket-ship of some kind at close to the speed of light. From twin 1's perspective, twin 2 has been speeding away at near light speed, so as we've established, his time must have slowed down, meaning less time has passed for him and twin 2 must now be younger than twin 1. All good, but what about from twin 2's perspective? For him, in his comfy rocket ship, it didn't feel like he was moving, he did however see the Earth with twin 1 on it receding quickly away out the rear-view window at something approaching the speed of light. From this perspective, twin 1's time must have slowed down, so twin 1 is younger. Both twins see the other twin as being younger, and both reference frames are correct. But how can this be? After all, this isn't the quantum world we're talking about, if they meet up again in the future, they can't be in a superposition of states being both older and younger at the same time, one must actually be younger. The simplified answer comes from which twin catches up to the other in order for them to meet up again. If for example twin 2 does a U-turn and heads back to Earth, twin 1 will be older. The key to take away here though is that both perspectives were correct. When they were separated, relative to each twin, the other twin was younger.

There are many examples out there of relativity-related mind benders like these two, but no matter how simple or complex, you'll never find one that actually lives up to the name of paradox. It turns out that the universe is perfectly capable of handling any scenario you can throw at it from any number of reference frames (no matter how confusing or contradictory they may appear at first).

## General relativity

10 years after his first publication on relativity (now known as special relativity), Einstein included gravity into the mix with general relativity. This is arguably what made him famous when its prediction of gravitational lensing (the bending of the path of light towards a gravitational body) was first confirmed during a solar eclipse.

General relativity reveals a relationship between gravity and time. A good example of this relationship would be that clocks at high altitude go faster than clocks at low altitude. Is there much of a difference? Well, just by standing up, walking around and generally going about your day, this effect causes your head to age more than your feet, but as you can probably guess, the effect is incredibly small (this difference is best measured in femtoseconds, a femtosecond being the amount of time it takes light to travel around 30cm or 1ft). This difference does have to be taken into account with various technologies though. GPS satellites for instance would lose ~10km of accuracy per day if they didn't take relativity into account.

## The arrow of time[1]

Why do we remember yesterday but not tomorrow? Why can we affect the future but not the past? Or more simply, why does the arrow of time point forwards? This might seem a straight forward enough question on the surface, but try to give a simple answer and you may find it becomes trickier than you think.

You may be surprised by the statement that there is no arrow of time to be found in the fundamental laws of physics. The rules don't differentiate between past and future. If one were to record a video of two billiard balls colliding and bouncing off each other, you'd be unable to determine from their interactions whether the video is playing forwards or backwards. The rules governing their interactions can be applied equally well regardless of which direction in time you're going. However, something different happens when recording a video of a cue ball breaking up a collection of racked balls. In reverse, the video would show all the balls coming together into a nice orderly pattern, and you could say with certainty which direction was forwards and which backwards on the video. With many moving parts, the arrow of time begins to emerge, so what's happening here? It comes down to entropy. Sticking with our billiard table analogy, when the balls are all nicely racked up, there are a very few possible configurations, and so the system has low entropy. When they're scattered about the place, there are a great many possible configurations, and so the system has high entropy. And you will never see a system go from high

---

[1]This section was heavily influenced by a talk from Sean Carroll.

entropy to low entropy. No one ever smacks the queue ball into a messy table to have it end up with all the balls nicely racked and set. The nature of time passing is that systems always tend towards being more chaotic, more disorderly, having a higher entropy, and this is true for any system. So to get back to the original question, tomorrow will come after today because tomorrow will have a higher entropy than today. The entropy we have today will have increased. Yesterday came before today, because we had lower entropy then. Don't worry, the universe is a long way from maximum entropy (called thermal equilibrium), so there are many more tomorrows yet to come.

With all this in mind, it may help to think of time much like space. Space has no intrinsic up or down direction, we only know which way is which due to our proximity to the Earth. Likewise, time has no intrinsic forward or backward direction, we only know which way is which due to our proximity to the big bang.

All this does lead to a new unanswered question. Today came after yesterday because yesterday had lower entropy, and the day before that had even lower entropy, and the day before that lower entropy still, and we can follow this process all the way back to the big bang, a point where the universe was unbelievably well ordered. But why was the universe ever in such a well-ordered state in the first place? We can certainly ascertain that it was, but as to why, while there are some hypotheses to explain it, they all still remain in the realm of speculation.

# 5. Space

## Vacuum energy

Space as it turns out, is an incredibly energetic and busy place. It may be easy to imagine the vacuum of space as completely empty, but what if you had two magnets, would you still expect them to attract each other in space? Magnets work by affecting the electro-magnetic field and this, along with various other fields (for instance, the gluon field) are always present, no matter where you are in space. Just because there are no magnets, doesn't mean the electro-magnetic field isn't still there doing its thing.

The presence of these fields cause various particle/anti-particle pairs to fluctuate in and out of existence, although thinking of them as actual particles is possibly slightly misleading, as they can do some very non-particle like things, like go in the opposite direction of their momenta. These appear so briefly before annihilating each other that they're referred to as virtual particles, it is however possible to disrupt this process. The idea was first put forward as a thought experiment involving a mirror rotating at near light speed, deflecting virtual photons away from their partners before they could annihilate. More recently, an actual experiment involving circuitry that effectively acts as a rotating mirror for microwaves was successful in getting some of these virtual photons to stick around as real photons.

So if we can get photons out of vacuum, does this mean that technology is on the way to getting energy out of empty space? Well, certainly not any time soon, if ever unfortunately. While space is technically extremely energetic, it's also already the lowest energy state. If you have a ball and drop it to your feet, you convert its potential energy into kinetic energy as it moves from having a higher potential energy to a lower one. If you had a ball high up on a plateau, your ball would have a higher potential energy, but dropping it to your feet would still result in the same amount of potential energy being converted to kinetic energy. In order to tap in to that higher potential energy, you would have to go to the edge of the plateau to drop the ball off. But what if the plateau went on forever? There would be no way to get at all that extra potential energy. Similarly for space, while there's a lot of potential energy there, to convert it to something we can use, we would have to find the equivalent of the plateau's edge, somewhere with a lower energy state, which, as far as we know, doesn't naturally exist.

## Black Holes

Stars are basically nuclear fusion reactors in space, continuously pumping energy outwards, but what happens when a particularly massive one, perhaps between 30-100 times the mass of our sun or so, reaches the end of its life cycle and runs out of fuel? Within milliseconds, the huge amount of the star's gravity causes the spent core to collapse in on itself, this gravitational collapse crushes it down so small that even atoms are crushed, it becomes a baby black hole, a singularity with such strong gravitational effects that time on its surface is infinitely slowed, and nothing, not even light can escape once it is close enough. This then quickly begins to cannibalise the rest of the star, resulting in a hypernova, an explosion so huge that in one second, 100 times more energy can be released than the total our sun will have produced by the end of its lifespan.

This kind of scale may seem difficult to come to grips with, but it doesn't stop there. After the birth of a black hole, and this is just one of multiple methods proposed, it can grow by absorbing whatever it encounters; stars, other black holes, pretty much anything. Unlike some sci-fi movies might lead you to believe, you really wouldn't want to get too close, you'd be spaghettified, and yes, that's the technical term, and yes, it's just like it sounds. When sufficiently large enough, they're considered supermassive black holes. It's expected that at the heart of nearly every galaxy of sufficient size there is a supermassive black hole (the supergiant galaxy A2261-BCG is a currently unexplained exception to this rule). The measurement scale used for black holes of sufficient mass is in solar masses. 1 solar mass = as much mass as our sun has. The largest supermassive black hole found so far is around 40 billion solar masses. This is at the heart of something called a quasar, which is believed to be the result of two galaxies colliding, allowing the supermassive black holes at their centre to merge.

So if nothing can escape a black hole, do they just continue to grow indefinitely and never stop? Well actually no, current theory expects them to 'evaporate' away given an extremely long period of time. Roughly a google years from now there should be none left (a google is a 1 followed by 100 zeroes). But how does this happen if even light can't escape? It comes back to the previously covered idea of virtual particle/anti-particle pairs. The event horizon of a black hole is the boundary, which once passed there's no coming back from. Now imagine a virtual particle pair coming into existence right on the edge of this boundary. It's possible that one falls into the event horizon and one doesn't, preventing their annihilation. From an outside perspective, the black hole has just emitted a particle, known as Hawking radiation, but energy is always conserved, so where did the energy for this new particle come from? The answer is from the black hole, which loses a proportional amount of mass (this loss is incredibly tiny when compared to the mass of a supermassive black hole, hence the enormous size of the timeframe for complete evaporation).

## Quasars

To say quasars are bright is an understatement. Only lasting for between 10 million and 100 million years, the galactic equivalent of a blink of the eye, they can be thousands of times brighter than the whole of the Milky Way galaxy combined. They're generally thought to be powered by two supermassive black holes orbiting each other, at the peak of a feeding frenzy, with the turmoil created around them resulting in the light that we see. Astronomers thought they had a good grasp on how quasars were formed, that is until 2015, when a surprising discovery somewhat threw that into question.

Quasars are understandably rare, normally being separated by hundreds of millions of light years. If you see two together in a small patch of sky, the chances are high that it's probably a single quasar being seen twice due to gravitational lensing, the path of the light being bent around something on the way to us, perhaps around a galaxy in between. It is possible to find quasars close to each other though, and to date, ~100 such quasar pairs have been found. There have even been two quasar triplets found. Three quasars in a small patch of space should be incredibly rare, but certainly not implausible. In 2015 however, a quasar quartet was discovered. The four quasars are within ~650 thousand light years of each other. As an idea of scale, our closest galactic neighbour, the Andromeda galaxy, is over 2.5 million light years away. If our existing models for the formation of quasars were 100% correct, then the odds of such a quartet existing and us happening across it would be incredibly tiny. It's overwhelmingly more likely that we're missing something in our theories, a piece of the puzzle which would perhaps allow quasars to form more readily under certain conditions. As of writing, while investigations are continuing, no widely accepted explanations have yet been put forward, leaving the quasar quartet as a physicist's favourite thing, an unsolved mystery.

## The Higgs field

One of the most well-covered scientific breakthroughs of recent times was the confirmation of the long sought-after Higgs boson, sometimes referred to in the media as the god particle (much to the dismay of many a physicist), but what actually was it? What was all the fuss about?

Every particle can be thought of as a condensation of the corresponding field, so for instance in the electromagnetic field, when there's enough of an excitation at a point, the condensation you get is an electron. The Higgs boson then is a condensation from the excitation of the Higgs field, a field which had been predicted by the standard model of particle physics (an extremely successful model), but that had never previously been confirmed. It was the confirmation of this Higgs field that was such a breakthrough.

Though the name doesn't suggest it, the speed of light is not unique to light; gravity for example also travels at the speed of light. Anything without mass cannot help but travel at this speed, so in fact mass could be described as the ability to travel slower than the speed of light, and this in turn is only possible due to the Higgs field. The Higgs field gives other things their mass through a process called the Higgs effect. You could think of it a little like a thick soup permeating all of space, where particles which interact with it slow down while passing through.

While that analogy may help with grasping the general idea, as you can probably guess, the process for giving mass is not quite so straight forward. For a more complete picture, firstly think about what happens when a massless photon of light encounters an object. It bounces off, reflected like a billiard ball bouncing off the side of a table. Secondly note that the Higgs field, compared to all the other fields, has a very high value at all points. Not enough that there are Higgs boson condensations around the place, but still very high. Now what happens to something which would otherwise be massless as it passes through this high value area? Just like the photon it gets reflected, bouncing back and forth, back and forth a great deal, but due to quantum mechanics, it's not bouncing back 'and then' forth, it's actually bouncing back 'and' forth at the same time. The speed in between these bounces is still the speed of light, but when everything is added up, it averages out so that the overall speed is slower than light, perhaps even close to stopped.

# 6. Misconceptions

## Tachyons

Thanks to countless sci-fi movies and the like, when you hear the word tachyon, you probably think of hypothetical, faster than light particles, as impossible to slow below light speed as it is to get normal particles above light speed due to their having negative mass. Well, while there's no evidence to support the existence of these (though that doesn't stop some people from looking), there is another, less well-known meaning for the word. In physics, tachyon can also refer to a tachyonic field which is a field with an 'imaginary' mass, something which counter-intuitively does exist, and is probably at least partly to blame for some of the confusion surrounding the tachyon particle's existence (or lack thereof).

The term imaginary here is borrowed from mathematics, where an imaginary number refers to a multiple of i, i being equal to the square root of negative 1. This allows mathematicians to play around with a number set which doesn't translate well to the real world (as you normally can't have a square root of a negative number). In physics, to ascertain the mass of a particle you need to square the associated field, and it's perfectly acceptable to have a field with an 'imaginary' value. So the question then becomes, if you take an excitation of a tachyonic field with imaginary mass, won't the resulting condensation be a particle with negative mass? After all, any imaginary number squared results in a negative number right?

Well, this would be true, except that due to instabilities created by the field's imaginary mass, when an excitation occurs, they spontaneously decay. Sometimes these can end with an alternate configuration with a positive value, but never with a negative one. An excellent example of this is actually the Higgs again. The Higgs field is tachyonic with an imaginary mass but, as established, the result of an excitation of the field is the Higgs boson, which has a definite, positive mass.

## Plutonium

Movies can get many things wrong. Sometimes these are deliberate, like adding sound in space to make a scene more exciting. Sometimes they're convenient, like having the sound of an incoming rocket for protagonists to react to before the explosion (for a rocket travelling faster than sound, the explosion would happen first, followed by the sound of the incoming rocket as the sound caught up). However, sometimes they stem from misunderstandings of the physics involved, and a good example of this is with plutonium.

Many people associate plutonium with great danger. You don't see a movie where someone has a block of the stuff as a paperweight, after all, it's radioactive, and you can make nuclear bombs out of it right? Well, technically both true, but that doesn't actually make it dangerous. In fact, the most dangerous thing someone could realistically do with a stick of plutonium is probably swing it at you like a bat. There are many different types of radiation, plutonium emits alpha radiation. The alpha particles it releases are identical to the nucleus of a helium atom, they don't travel very far through air, and are completely unable to even penetrate skin (the layer of dead cells found on the outside of skin is more than enough to completely stop them in their tracks). Theoretically, if it was broken down into microscopic components of the right size and inhaled, the radiation could affect someone through the lungs, which don't have such a protective layer of dead cells, but even if you were to explode some plutonium with dynamite, you wouldn't get particles of the correct size for this. But wait, what's this talk about exploding it, surely it should at least be handled with care right? What about the bomb side of things? Well there's another misunderstanding here. There are different types of nuclear bomb, but the type that employs plutonium requires an implosion. That is, you need to rapidly compress the plutonium from all sides equally for anything to happen. Ever tried squeezing a water balloon? You squeeze a little on this side, and it expands out over the other side. Try compressing it equally on all sides and you'll struggle. Now just imagine trying to do that to plutonium using explosives, and you'll have some idea of how incredibly difficult it is to actually make a bomb utilizing plutonium.

## The Big Bang

Names can sometimes be misleading. When first introduced, Nuclear Magnetic Resonance (NMR) was refused by dying patients due to a fear of the word nuclear, associating it with nukes, even though it was referring to detecting hydrogen nuclei in the presence of a magnetic field, and not something to do with radioactive materials. This trouble disappeared after it was renamed Magnetic Resonance Imaging (MRI). However, possibly the most misleading name in all of physics is the 'big bang'.

From the name, many take away an idea of the universe exploding into existence from a point with a bang, but an explosion would require there to be space to explode into, and this imagining of it from an outside perspective is generally where any major misconceptions come from. The trick to getting a better picture is to understand that as the universe expands, it's not actually expanding 'into' anything. Normally if something becomes bigger, it takes up more space, however by definition, the universe contains all of space. There is no space outside the universe to take up more of, and so when the universe expands, it doesn't take up more space, it instead simply contains more space. If it helps, you could think of the big bang as the point where the expansion and cooling of space began. Certainly not as catchy as far as names go, but possibly less misleading.

## Quantum tunnelling

Anywhere from book covers to super-hero insignias, you can find drawings of the classical physics picture for what an atom looks like. They have the nucleus in the middle made of protons and neutrons, while on the outside there are some electrons whizzing around in circular orbits. This iconic picture for physics may be simple and catchy, but it's far from accurate, and likely misleading for those who use it as their go-to visual for understanding the workings of the atom. For starters, there's the sense of scale. The nucleus of an atom is tiny, generally around 100 thousand times smaller than the rest of the atom. If you were to imagine the atom represented by a football stadium, then the nucleus of the atom would be represented by something the size of a mosquito in the middle of the stadium. There are other things too, like the idea of Bohr's orbits (the electron going around the atom in a circle). This idea has long been superseded by that of probabilistic electron clouds. The electron doesn't 'move' at all. If you were to draw this, the electron around the atom would resemble more of a fog, with denser regions of the fog corresponding to more likely places to find the electron at any given time. Many might be surprised to discover that it's even possible to sometimes find electrons hanging around inside an atom's nucleus, certainly not something that the classical picture would ever lead you to predict.

This again comes back to the probabilistic nature of quantum mechanics. The electron's default setting, being governed by the rules of quantum mechanics, is in the previously discussed state of superposition. The wave function that can be used to describe an electron can extend beyond a barrier that it normally couldn't pass, and this wave-function represents the probability of where you'll find the electron. So, when you choose to observe where the electron actually is, while you're still likely to find it on the side of the barrier you'd expect, where the majority of the wave function is, there's a small probability that you'll find it on the other side of the barrier. This is referred to as quantum tunnelling as the electron appears to have 'tunnelled' through an otherwise impassable barrier. There's even a small chance that you'll find it inside the barrier, hence being able to find an electron hanging around inside an atom's nucleus. Quantum tunnelling was the key concept behind the 1986 Nobel Prize in physics for the design of the scanning tunnelling microscope, a device that allowed us to see details down to the nanometer scale. This remained probably the largest advance in imaging until the development of what is commonly referred to as nanoscopy, which won the 2014 Nobel Prize in chemistry. Physicists aren't the only ones who want to study the small scale in as much detail as possible.

# 7. Energy

## Nuclear fusion

While nuclear fission and nuclear fusion may sound similar, they are in fact nearly exact opposites. Nuclear power plants (and bombs) rely on nuclear fission, harnessing energy released from the splitting apart of an atom. Nuclear fusion on the other hand, has sparked interest as a potential technology of the future, obtaining energy from the process of joining (or fusing, hence the name) of atoms together.

The basics are simple enough. At the very small scale, you can imagine that if shuffled together, all protons and neutrons would prefer to come out as part of a helium nucleus. If you can find a way to facilitate the components of hydrogen or helium isotopes to do this (usually the positive charge from the protons in atomic nuclei prevent them coming close enough to each other for such a shuffling to occur), then any leftovers after the reaction will fly out and can be converted to energy. There are several approaches to this, all of which currently have their own drawbacks. We'll quickly look at just three.

ITER is probably the most well-known of these attempts, an international collaboration in southern France attempting to achieve fusion through magnetically confining a plasma at extremely high temperatures. When hot enough, natural thermal motion brings nuclei close enough together that nuclear forces (the force that holds atoms together) can begin to dominate. While many obstacles have already been overcome, ITER's greatest challenge is likely to be that their reaction results in energetic neutrons. Neutrons have no charge, so in order to get energy out of them, something must be put in their way, letting them heat it up through bombardment, but this deteriorates whatever it is that's being bombarded extremely quickly. At the moment, there's no substance known which can act as this shielding, usefully coping with the sustained output of such a reaction over the long term while also allowing energy to be extracted.

On the other side of the planet in California is the National Ignition Facility (NIF). This facility, roughly the size of three football fields, is basically filled with high-powered lasers focused on a single point, the idea being to extremely rapidly heat the outside layers of a fuel pellet, compressing it so that the density and temperature at the core become enough that a fusion reaction begins. While this approach has also solved many obstacles along the way, their greatest challenge was probably in attempting to break even. That is that it takes more energy put into the high-powered lasers than is obtained from the resulting fusion reaction. Because of this, the facility has been focused on other areas of research for the last few years.

The third and possibly least well-known approach is muon-catalysed fusion, otherwise known as cold fusion due to its lack of reliance on temperature as a catalyst. Naturally occurring muons travel extremely fast, there's probably around 1 per second passing through the palm of your hand if you hold it out right now. Slow

43

moving ones can be artificially created though, and you could imagine a slow moving, negatively charged muon much like being a big, fat, short-lived electron. If you introduce one into a mixture of hydrogen isotopes, it will take the place of an electron around a nucleus, but due to its size, it orbits extremely closely to the nucleus. So closely that it essentially cancels out the charge of the proton for any neighbouring atoms. The chargeless nucleus can then wander until randomly close enough to another that the nuclear forces take over, resulting in fusion, with the muon generally being released, able to go catalysing further reactions. Mid last century the greatest problem with this approach was the catalysing lifespan of the muon. A single muon just couldn't catalyse enough reactions to be energy efficient. This problem (along with others) was solved more recently using an isotope of helium (helium-3) as the fuel, but replaced with a different problem. Current muon generation techniques are not sufficient to reach the saturation level needed, a flux a few orders of magnitude above what we're currently capable of is still required.

So with such a varied array of problems for different approaches, is fusion ever likely to be used at all? Well, while there is a bit of a joke along the lines that no matter how much time passes, fusion technology always seems to be around 20 years away, progress is still being made, and while there may still be any number of unforseen setbacks in development, it's probably still a fairly safe bet that it will eventually find its place as an important technology in mankind's future.

# Nuclear fission

Break apart an atom of fissile fuel with a neutron, and you get some energy, 2 fission fragments, and a few neutrons released. These neutrons if they encounter more fissile fuel will break it up with the same result, releasing more neutrons and so it can go on. This is why it's called a chain reaction. Unlike with nuclear weapons, nuclear power plants control these fission chain reactions by using a neutron moderator, a liquid, generally water, to absorb some of the neutrons released. It's a balancing act, absorb too many neutrons and the chain reaction will stop, but absorb too few and you risk a meltdown from releasing too much energy too quickly. The statistics needed for this juggling act are well understood, but what might surprise you is that mankind wasn't the first to get this happening, nature beat us to it by nearly 2 billion years.

1.7 billion years ago under what is now Gabon in Africa, a uranium rich mineral deposit was flooded with groundwater, and the conditions were just right to form the only known example of a naturally occurring nuclear fission reactor. The heat from nuclear fission would boil the water away, things would cool, water would return and the cycle would begin again. This ~3-hour cycle repeated for hundreds of thousands of years, continuing until there was not enough fissile material left. It has been pointed out that now, almost 2 billion years later, by-products of this reaction have yet to move more than a few centimetres, a detail which has been used as a natural analogue for nuclear waste disposal.

## Conservation of energy[2]

What is energy? We know it can take various forms, chemical, elastic, electrical, gravitational, heat, kinetic, mass, nuclear, radiant, but what actually is it? When attempting to answer this, Feynman used an analogy of a child with a set of indestructible blocks. At the end of the day, the mother could come in and count the blocks, always finding their number to be the same. If she found less blocks, with further looking she would perhaps find the missing ones hidden under the rug or thrown out the window. If they were hiding in places she couldn't look, like in a locked chest, she could weigh the chest, deduct its normal weight, and deduce the number of blocks inside, noting that the amount calculated corresponded exactly to the missing quantity. Ways of calculating block numbers in places she couldn't look could get more complicated, but the rule clearly emerges that their number must always stay the same. He abstracted from this picture an analogy to the law of conservation of energy. Take away the blocks from the equations, and you find yourself calculating essentially abstract things. It is these calculations that we can apply to energy. Much like being hidden from the mother in a locked chest, energy can change forms so that we need different techniques to measure it, but after all the calculations are completed, we will find that energy has not been created or destroyed, we will always end with the same amount that we started with. So back to the original question, what actually is energy? Well we really can't say what it is in the usual sense. We

---

[2]Sometimes, an existing summary already exists that is so good that any alternative feels inferior. This is one of those times. As such, the following section draws heavily on the writings of Richard Feynman.

can calculate some numerical quantity which never changes whatever is done to it, and this is what we refer to as energy, but we can't actually give a reason for the formulas, all we can do is ascertain what they are.

## Negative Energy

In some theories, the existence of sufficient negative energy could be used for everything from keeping wormholes open to building a warp drive. Unfortunately that kind of negative energy is currently confined to the realm of the theoretical, but negative energy is still a very real concept.

As the vacuum of space is the lowest natural energy state, it's reasonable to see it as a zero point, but as previously mentioned in the section of vacuum energy, it's actually still extremely energetic. So what happens if we somehow artificially lower the energy of a region below this zero point? Then technically that region would count as having negative energy.

Probably the most famous example of this is the Casimir effect, named after Hendrik Casimir who came up with the experiment roughly 50 years before technology was able to put it to the test. Imagine a vacuum as being filled with energetic waves corresponding to their various fields. These waves are energetic, and so do exert a force, but as this force is everywhere and identical from all sides, it's never normally noticed. Now take two plates and place them nanometers apart from each other. This limits the waves between the plates to having certain frequencies (recall that in quantum mechanics, you can't have an incomplete wave). With more waves on the outside of the plates than there are between them, a force becomes noticeable, attempting to push the plates together. The closer the plates, the more wave frequencies are prevented from existing between the plates, the greater the difference between the energy outside the plates compared to between the plates and the greater the force pushing them together that's experienced. The region between the plates has less energy than the vacuum of space, it has a negative energy density.

In recent times this force has been studied in some detail, with some very different ways of explaining it suggested. However, as of yet, suggested practical uses for it have been quite slow to emerge.

## $E=mc^2$

Energy equals mass times a constant (the speed of light) squared, or $E=mc^2$. Probably the most quoted piece of physics in the world, but what you may not have realised is that it's not the whole equation. Take light for example, which has zero mass. Plug that into the equation without any changes and you'd get that it has zero energy, something's missing.

$E=mc^2$ is for particles with mass, which aren't moving. If you wanted the equation for particles without mass, like light, then it would be $E=pc$, where p is a particle's momentum. But again, this is another incomplete formula for a specific circumstance. What if you wanted an overall equation for all occasions though?

You may think that this would require getting quite technical and maths heavy, but actually it doesn't take any more than the Pythagorean theorem you probably encountered in school. For a quick refresher, if you draw a right-angle triangle (a triangle where one of the corners is 90°), then the length of the hypotenuse, the edge opposite the 90° corner, can always be figured out given the length of the other two edges with another rather famous equation $a^2+b^2=c^2$. But how does this help with the equation for energy? Well, let one side of the triangle, side a, deal with mass. Let the other side of the triangle, side b, deal with momentum, and let the hypotenuse, side c, be the energy that we're trying to figure out. Now we can simply plug in our two circumstance specific equations using Pythagoras' $c^2=a^2+b^2$ and we get $E^2=(mc^2)^2 + (pc)^2$.

So why not just give this equation to start with, why all the fiddling around with Pythagoras' theorem? Well, if you can visualise the three sides of the triangle, where one is $mc^2$, one is pc, and the hypotenuse is E, then it becomes possible to easily appreciate certain things which may previously have been more of a mystery. Take the speed of light for example. If you want something with mass to go at the speed of light, simply try drawing the right-angle triangle such that E, the hypotenuse, is the same length as the momentum side (pc). It should quickly become apparent that this can't be done, as no matter how much you fiddle, the hypotenuse is always going to be the longest side of the triangle. Now you don't just have to take a physicist's word that getting something with mass up to light speed can't be done, you can see it for yourself just by drawing a right-angle triangle.

# 8. The Universe

## Dark matter/energy

What makes up the bulk of the universe? While there's plenty of hydrogen and helium gas out there in space, this actually only accounts for ~4% of the universe. The generally accepted answer is dark matter and dark energy. If current theories are correct, dark matter takes up an impressive ~25% of the universe's mass-energy, and is believed responsible for certain gravitational effects, but that's nothing compared to dark energy which takes up a whopping ~70%, and is believed responsible for the accelerated expansion of the universe. Interestingly, neither is fully understood as they were both relatively recently theorised in order to explain otherwise anomalous effects.

When looking at a distant galaxy cluster, it's possible to estimate how much mass it has based on its brightness and number of galaxies. You can also estimate its mass based on the gravitational effect it has on the motion of neighbouring galaxies. What do you do if the second figure is substantially more than the first though? Infer there must be something else there, give it a name (dark matter, as whatever it is it's non-luminous), and start trying to understand it. Dark energy has a similar story. The universe is expanding, for a long time there was some speculation about whether this would continue indefinitely or whether gravity would cause it to start collapsing in on itself again (nicknamed the big crunch). This big crunch option was pretty much dismissed with the surprising discovery that the expansion of the universe was not slowing down but accelerating. What would explain this unexpected discovery though? There must be a previously unknown force at work, give it a name (dark energy), and start trying to understand what it is as well.

Dark energy can probably be most easily visualised as a constant energy density evenly filling all of space. If there's a volume of empty space, it'll have a certain amount of dark energy. Dark matter on the other hand has a bit more structure to it. Imagine around every galaxy you have a clump (or halo) of dark matter, then imagine these being connected, spanning the distance between galaxies with thin strands, known as filaments. It's much like a 3d spider's web, only invisible, and on a universal scale.

## Size of the universe

It's not uncommon to find someone saying that people used to 'know' the Earth was flat, and they were wrong, so how can we be sure we're right about anything we 'know' today? But theories are never so much wrong as incomplete. Once scientists get hold of a good concept, they gradually refine and extend it with greater and greater subtlety as their instruments of measurement improve. Yes, you could say that we were wrong then, and yes, you could say that we might be wrong now, but to say that we might be wrong in the same sense as we were then is, to borrow from Asimov[3], "wronger than both of them put together". With the tools available to ancient humans, a curvature of 0, or a flat Earth, was a perfectly reasonable deduction. With more precise measurements, we can say that if the Earth were completely spherical, it would have a curvature of ~0.000126 per mile, but the Earth is not completely spherical. Due to its rotation it bulges around the equator making it more of an oblate spheroid. While a correction from sphere to oblate spheroid has much less impact than a correction from flat to spherical, it's still just another correction from the data available, and we can continue further still. If we look at millionths of an inch per mile which modern technology allows, we can say that the South Pole sea level is ever so slightly different to the North Pole sea level, so in fact the Earth could be called more of a pear shape than an oblate spheroid. If you were feeling particularly argumentative, you could even put forward the case that the world is indeed flat, it's just a matter of perspective. Remember Lorentz contraction in special relativity? If a cosmic ray were approaching the Earth at high enough speeds, from its perspective the Earth would be contracted into a pancake shape. The point is that while corrections to our knowledge are likely, and indeed expected in the future, that doesn't make our previous or current theories 'wrong', merely incomplete to varying degrees. This may all seem relevant when discussing what we know about the size of the universe.

---

[3]Much of this paragraph was heavily influenced by writings from Isaac Asimov.

Current measurements show the cosmological curvature parameter (the curvature of the universe) to be 0.000±0.005, which is consistent with the idea of a flat universe. However, we can only measure as far as we can see into the universe, a region called the observable universe, which has a radius of approximately 93 billion light years (or 28.5 gigaparsecs), and currently have no way of really telling what it's like beyond this, somewhat limiting our ability to make solid determinations. Perhaps the universe has many bumpy regions and we just happen to be located in a section that's flat, like standing in the middle of a huge plain, surrounded by mountains too distant to see. Perhaps it's much like the Earth was for ancient humans, generally consistent on average throughout, but with the slightest of curvatures, undetectable with current technology, but allowing the universe to essentially be spherical on a truly enormous scale. There are other possibilities, but even if the entire universe is completely flat, space doesn't have to follow the normal rules of geometry, it could still be what's known as multiply connected, allowing it to actually be flat and finite at the same time. So to come back to the original question, what's the size of the universe? Well, currently we just don't have enough data to say, we're much like the ancient humans were when trying to determine whether or not the Earth was flat. The universe certainly looks flat and infinite with the tools currently available to us, but it's far too early to say for sure. But even should we get enough data to strongly support the idea of finite or infinite at some point in the future, the theory like any other will still be open to being refined and extended. Remember, with this theory or any other, don't be sold by the logic that we're ever going from being wrong to being right, theories are based on the data available, and so only ever go from being less refined to more refined as more data is accumulated to work from.


## Energy of the universe

The universe is expanding, taking a little over 3 billion years to double in volume at the current rate. Additionally, the vacuum of space has an energy density associated with it. Combine these two facts and a question arises. How can these two things both be correct given the law of conservation of energy? Wouldn't an expanding universe with a constant vacuum energy density lead to there being more energy in the universe as time goes on?

There have actually been a few different attempts to address this apparent paradox. One is to start talking about a different type of space-time, so there's Minkowski space-time which treats space as a flat backdrop and covers all the day to day regular stuff, where things like the law of conservation of energy are set in stone, and then there's FLRW (Friedmann–Lemaître–Robertson–Walker) space-time, which deals with space-time on the grandest of scales. Here the law of conservation of energy is not enforced, allowing the overall energy of the universe to indeed increase over time. Another approach says that over time, the energy is staying constant, but being spread out as the universe expands. This leads to a prediction of an energy drop in the vacuum energy density over time. However, after calculations are done, the predicted drop is so small as to be well below what current technology is capable of detecting and so while possibly testable eventually, confirmation remains well out of reach for the foreseeable future. The approach which is probably the most fun to speculate on makes the assumption that the universe is closed, that space eventually bends back on itself. Given this assumption, it's then possible to assign a negative energy to the curvature of space. So given any size universe, yes there's energy in it, but this is always exactly balanced out by the negative energy from the space-time curvature. The bigger the universe, the more energy within it, but the more negative energy from the space-time curvature, meaning that regardless of size, the total energy of the universe is always exactly zero.

## Supervoids

When thinking of the biggest structures in the universe, you may think of galaxies, in which case the supergiant elliptical galaxy A2029-BCG (also known as IC 1101) probably holds the title, with a diameter estimated at roughly 60 times that of the Milky Way, and a mass of ~100 trillion stars compared to the mere ~100 billion of the Milky Way. You can have walls of galaxies, the most well-known of which is probably the Sloan Great Wall (SGW). While possibly a chance alignment of three individual structures, it still stretches an impressive ~1.38 billion light years. There are other things which have scales equally as difficult to come to grips with, but one possibility that can be easily overlooked when thinking of biggest structures in the universe is that of voids, and a good example of this is a supervoid known as the Boötes void.

The Boötes void is a roughly spherical section of space with a diameter of ~330 million light years (~0.27% of the diameter of the entire observable universe). Voids aren't completely empty, they're just extremely underdense regions of space. If galaxies were distributed evenly throughout space, then you would expect a region of space the size of the Boötes void to contain perhaps ~2000 galaxies, but instead, just 60 have been found, forming a roughly tube-shaped region through the middle of it. One astronomer[4] stated that "if the Milky Way had been in the centre of the Boötes void, we wouldn't have known there were other galaxies until the 1960s". Current theory suggests that it's a collection of smaller voids which have come together over time, eventually coalescing into the one supervoid. An analogy which is sometimes used is that of soap bubbles coming together to form one larger bubble. This theory helps to explain the tube of galaxies running through the middle.

The Boötes void isn't the only supervoid out there, with the largest one confirmed so far dwarfing it in size. Known as the Canes Venatici Supervoid, or more simply, the Giant Void, it's ~4 times the size of the Boötes void with a diameter of almost ~1.3 billion light years, nearly the same as the SGW. One amusing thing to contemplate when considering voids is the possibility that perhaps, just perhaps, the entire observable universe is itself part of an underdense region, a void on a truly grand scale. If this were the case, then there would be no need for dark energy to explain the accelerating expansion of the universe, it would merely be the result of our being in the middle of a huge underdense region, unable to see what lay beyond, much like if the Milky Way were in the centre

---

[4]Gregory Aldering

of the Boötes void, we probably would have come up with all sorts of dark galaxy theories over the years to explain why there were no other observable galaxies. While the chances of dark energy being behind the accelerated expansion of the universe are extremely high, until better understood, the void idea remains, for the time being at least, a plausible, if remote possibility.

## Eras

When are we? A question that depends heavily on what kind of time frame you're interested in talking about. Checking a calendar is only relevant for modern times. If you want to discuss where in human history we are, it's likely convenient to refer to ages (iron/bronze/stone etc). When talking about the Earth's four and a half billion odd year history, the timeframe again changes, but it can still be divided into convenient periods to talk about (Cretaceous/Jurassic/Triassic etc), so what about when the timeframe is changed to be universal, what then?

The universe is currently ~13.8 billion years old, but over the course of its entire existence it's expected to go through roughly five eras. The first was the primordial era, lasting from the big bang until stars and galaxies started popping up, this was the start of the stelliferous era which we are still in today. Eventually though all the stars will exhaust their fuel. The brightest ones will run out first, so for a time the universe will consist of some white and brown dwarfs, some neutron stars and some black holes, this will be the degenerate era, which is like a stepping stone to the following rather long lasting black hole era. As the name implies, during this time the only things of any note will be black holes, but as mentioned before, given long enough every black hole will eventually 'evaporate', and when none are left, the universe will find itself permanently in the dark era.

Are these the only possible eras? Not necessarily actually. It's always possible, or even likely, that there's something that we just don't know enough about yet. A good example of this might be the Higgs field. Mathematically speaking, it may be possible for the Higgs to have two distinct states, the one we know today, and another, ultra-dense state. If the ultra-dense state is a real possibility, then the possibility also exists that at some time, somewhere in the universe, a phase shift might occur, creating a bubble of this ultra-dense Higgs. It would then expand out at the speed of light, wreaking havoc on any interacting particles in its path as those particles would substantially gain in mass. Don't worry, you're extremely unlikely to suddenly be wiped out by an expanding ultra-dense Higgs field. Even if it is a real possibility, given a google years the chance would still be slim of such a phase shift occurring anywhere. Given an infinite amount of time though, like that which the dark era would otherwise last for, and the fact that it would only need to happen anywhere once, perhaps there could be another era in the universe's future, one where the ultra-dense Higgs dominates.

## Expanding space

There is one question that gets asked a lot given a very basic understanding of physics. It may have crossed your mind while reading a different section; unfortunately though, many attempted answers fail to address it correctly. The logic goes that if space is expanding, then galaxies are getting further away from each other, so galaxy A is getting further away from galaxy B, galaxy B is getting further away from galaxy C etc. If you follow this far enough, which the universe has plenty of space for, you'll find two galaxies with so much expanding space between them that they are getting further away from each other at a rate greater than the speed of light. But how can this be if nothing can travel faster than the speed of light?

The answer is likely not what you suspect. Yes, the distance between the galaxies is increasing with time, but no, the galaxies themselves are not actually moving, it's the space between them which is expanding, and this technicality allows galaxies to be treated as stationary. Ok, so there is 'some' galactic movement, for instance in around 4 billion years our Milky Way galaxy is going to collide with the Andromeda galaxy, not to mention that all galaxies spin, but this isn't movement due to the expansion of space. With regards to the question of expanding space, galaxies can be seen as being located at fixed points, and if something isn't moving through space, it certainly isn't travelling through space faster than the speed of light.

# 9. Spin

## Tennis racket theorem

Something doesn't have to be going anywhere to be in motion, it can be spinning. Chances are you're already familiar with the basic physics of this if you've ever spun a coin on a table or played with a spinning top. There are plenty of terms used to describe what happens, but the simplest way of thinking about it is probably, if not spinning then a top will fall down. If spinning, then it will 'fall' sideways. This is described as precession, but there's a lot more to spin than just keeping something upright.

Next time you have access to a tennis racket, try this. Hold it by the handle with the head out flat and find a way to identify the difference between it facing up or down. There might be a brand name on the handle that's either right way up or upside down. If not, perhaps you can draw an arrow on. Next, throw the racket up so that it rotates a full 360 degrees, head over handle followed by handle over head, and catch it. The tennis racket has done one full rotation, so you might expect that it will be back to where it started, with the brand name or arrow facing the same direction as before, but chances are high that it will now be opposite. Try this multiple times and you'll find that it's actually very difficult to rotate a racket in this way without it also doing an additional flip in the process. This is because the axis of rotation we're attempting to rotate the racket along is unstable, so it only takes the slightest of disturbances to cause a flip. While this is generally referred to as the tennis racket theorem, you could attempt it with any similarly shaped object, a TV remote or even an elongated phone, just don't go blaming this book if you end up dropping them while playing around.

## The Magnus effect

If you wanted to make the longest shot possible with a basketball, assuming you had as many tries as you liked so accuracy wasn't an issue, how would you do it? You could take the shot from up high so that the ball could go farther, but you'd still be throwing the ball in an arc, which means that eventually extra height would stop making any difference. Normally the ball's horizontal movement is slowed by air resistance until at the end of the arc it's just travelling straight down. From a sufficient height though, if you were to drop the ball but with a lot of backspin, you'd see the ball follow a very different flight path. It would start picking up horizontal movement as it fell, shooting off in the forward direction. This would be making use of something known as the Magnus effect.

There are lots of ways of explaining what's happening, but instead of the common approach of talking about pressure, the easiest way of understanding the effect is probably just to think of Newton's third law, that for every action, there's an equal and opposite reaction. As the back-spinning ball falls, the air rushing past the front of the ball is somewhat dragged up and over the top, while the air rushing past the back of the ball is being held back a little. Overall then, the ball is applying a backwards force to the air, so the air in turn applies a forwards force to the ball. Many sports make use of this effect, like spin shots in table tennis, or golf, where the ball is dimpled to help with grabbing the air as it flies, and the club is angled backwards so that backspin is added to the ball when hit. These factors combine to greatly exploit the Magnus effect allowing the ball to fly higher and farther than it otherwise would. The Magnus effect was also famously used in a second world war bombing run, later to be known as 'Dam Busters', where a backwards spinning bomb dropped from a specific height was used to cause an explosion in a place which otherwise couldn't have been achieved.

## Rattlebacks

At some level many physicists are much like big kids who love playing with their toys. Whether that's working with particle accelerators, choosing to use their laser pointer for absolutely anything, or just fidgeting with some permanent magnets, there's a wide array of possibilities. Probably the oldest physics toy around though is the rattleback.

Rattlebacks date back thousands of years, have gone by many names in many different cultures, been made in various sizes with materials ranging from wood to jade, and depending on their exact design can have slightly differing properties, though the basic idea remains the same. Nowadays, the most common by far are made of plastic and are about the size of a finger. So what qualifies a lump of plastic with a rounded underside and no moving parts as a physics toy? Well, if you place one on a smooth surface like a tabletop and try giving it a spin, something rather surprising can happen. For most rattlebacks, if spun in one direction it just spins until it stops, no surprises there, but if spun in the other direction, it will quickly begin to wobble up and down (or rattle), and then actually reverse its direction of spin, hence the name rattleback. At first glance this seems to go against common sense, or more specifically, seems to violate the conservation of angular momentum law of physics, so what's actually happening?

The key to understanding this property of the rattleback is in the asymmetric distribution of its mass. There are three axes to consider when spinning anything, and much like the tennis racket theorem, when spinning along one, another can become unstable. When spinning the rattleback on a table, instabilities are actually growing in both the other two axes (pitch and roll), and depending on which direction the rattleback is being spun depends on which of these two instabilities dominates. If pitching instabilities dominate, they quickly build to the point that causes the rattling and reversal of spin direction. If rolling instabilities dominate, while this would also eventually result in a similar rattle and reversal of spin direction, the growth of these instabilities happen much more slowly, allowing friction to bring everything to a stop well beforehand.

## Discs

Ever wonder why a solar system or a galaxy is mostly flat? It's the same reason that planetary rings, or even black hole accretion discs are also always flat, something called the conservation of angular momentum. If you take a cloud of particles all moving in random directions, like our solar system was made up of originally, while it would be incredibly difficult to work out the path of an individual particle, mathematically they can be described as having a single, total amount of momentum that they spin around their centre of mass. It may be near impossible to figure out what plane this spin is in initially, but give it time and it will become clear as collisions between up and down moving particles relative to this plane cancel each other out, leaving behind the familiar spinning disc shape.

## Particle spin

The story of spin has one more level to delve into. Moving electrons create a magnetic force; the faster they move, the greater the resulting force. So how do permanent magnets work? The electrons aren't going anywhere if the magnet's just sitting still right? Well, remember that something doesn't have to be going anywhere to be in motion, it can be spinning. If the electron were spinning then the electric charge could be seen as moving in a circle, and that would make for a magnetic force. The more electrons lined up spinning the same way in an object, the greater the force would be and the stronger the magnet.

This idea can be a useful way of understanding the basics of electron spin (in fact pretty much everything in the quantum world can be categorised by its 'spin' value), but beyond this analogy, the term 'spin' can become more misleading than useful. Electron spin never changes, and only has two possible orientations, referred to as spin up and spin down. This is due to the electron not being (so far as we are able to tell) composed of any smaller components. Electrons are not little tiny ball-like objects in the quantum world, they're point like particles, which means they do not have a surface area at all. They therefore do not 'spin' in anything like the same sense that objects in the macroscopic world we're used to do, they go directly from spin up to spin down, not passing through any in between positions. However, despite its possibly misleading nature, the term 'spin' has stuck.

Particles with different spin values follow different rules. A spin 0 particle like the Higgs boson can only have a single quantum state. You could imagine it as a perfect sphere where it looks the same no matter how you rotate it. However, rotate an electron, or any particle with the same spin value (electrons have a spin value of ½) by 360 degrees and it's in the opposite state to where it started from. You have to rotate it through a second full 360 degrees to get back to your starting point. This is not just a likely occurrence as it was with the tennis racket theorem, it must happen every time. Conveniently, Dirac managed to give an analogy, known to some as the plate trick, which helps give an idea of what's happening here rather well. Hold your hand out with your palm face up. Keep your palm facing up and begin rotating your palm by bringing your hand in towards your stomach. Keeping your palm facing upwards, continue the motion as far as you can by bringing your elbow out and letting your hand continue underneath your shoulder. Once you can go no further, extend your arm and you should now be in a rather uncomfortable position with your arm out to your side. Your palm is still facing upwards, but overall it probably looks like an invisible bouncer has a hold of you. Keeping your arm as straight as you can and keeping your palm facing upwards, bring your entire arm around, swinging it over your head in a large circle and you should find that you can continue until your arm is once again in front of you in the starting position. Once you understand the motions, you can see that if you were to balance a cup and saucer on your palm, the cup would rotate 360 degrees twice. After one rotation, the cup's handle will be facing the same way as when it started, but you'll be in a somewhat twisted position. After the cup rotates another 360 degrees, you're back to where you started. It might feel like a particularly messy dance move, but no one ever claimed imitating a quantum particle would be elegant.

# 10. Everyday things

## Pendulums

For hundreds of years, the regular motion of a pendulum was used for timekeeping, old grandfather clocks being a classic example. An ideal pendulum, free from things like friction and air resistance could in theory keep perfect time, however in practice reality gets in the way. For instance in the 18th century one of the most significant problems was the expansion and contraction of the pendulum rod with temperature, meaning that clocks would run substantially slower in summer. Today, while no longer used for timekeeping, pendulums are still useful in other areas, like being incorporated into many skyscraper designs to help minimise earthquake damage. This exploits the fact that they automatically sway counter to the motion of the earthquake. With a heavy weight at the end, this applies a force in the opposite direction, thereby suppressing the overall motion caused by the earthquake.

On the mathematical side of things, the double pendulum, with one pendulum hanging from another, is a great example of a chaotic system when studying chaos theory, highlighting the same principle as with the more well-known butterfly effect, where if you had perfect knowledge of the present, you could in theory predict the future, but even the slightest uncertainty leads to wildly unpredictable outcomes. However, from a mostly physics perspective, possibly the most interesting thing done with a pendulum to date has been with the Foucault pendulum.

A Foucault pendulum is just a long pendulum, but instead of being constrained to swing from side to side like a grandfather clock, it's free to swing in any direction. A pendulum will continue to swing the same way, regardless of the motion of its pivot. The pivot, where the top of the pendulum is attached, is in this case connected, via supporting struts, to the Earth, and so the result is that as the Earth rotates, the Foucault pendulum will appear to precess over the course of a day or more, rotating through a full 360 degrees (the exact amount of time for this depends on your latitude). The Foucault pendulum was the first ever experiment to be able to demonstrate the rotation of the Earth without relying on astronomy.


## Surface tension

Just through day to day interactions with the world, many rules of physics are intuitively understood, even if the behind the scenes workings generally aren't explored. For example, if someone were about to blow detergent bubbles at a child's birthday party, you would instinctively know in advance that a single bubble floating through the air will be spherical, but possibly not realise that this is a consequence of surface tension, the liquid composing the bubble attempting to find the shape with the least surface area, in this case a sphere.

Surface tension is the result of there being a greater attraction between liquid molecules than there is between liquid and air molecules. This causes an inward force at a liquid's surface, resulting in a kind of elastic membrane, easily seen by carefully filling a glass of water to slightly above the top. If filled slowly the water can easily go slightly higher than the top of the glass before gravity dominates, causing it to run down the sides.

The line of questioning that comes up most when talking about surface tension predictably revolves around a person falling a great distance without a parachute into water, so to quickly address this, yes, the surface tension of the water encountered at high speeds will be like concrete to the person falling, but no, they probably can't just drop/throw/shoot something into the water before hitting it to break the surface tension. Ok, technically they can, but if they're falling from a great enough height, it's still not going to save them. Even 'if' their hypothetical object has perfect timing, breaks the water's surface tension, gets out of the way before the person hits, and the water is deep enough so that they don't hit the bottom, the real problem is one of sudden deceleration. There's a big difference in speed between falling through the air (at maximum speed, or terminal velocity) and falling through water, and the sudden change from one speed to the other isn't going to be overcome by breaking the surface tension before impact. Of course, a bad plan is arguably better than no plan at all in such a situation, but the person falling is still going to have to hope that the water somehow becomes suddenly aerated (which would lower its density) or the like before they reach it, and normally in the hypothetical scenario, the absent-minded parachutist didn't push a few tons of Alka-Seltzer out of the plane before jumping.

## Ice skating

Get a physicist to explain enough things to you, and it probably won't be too long before an analogy involving ice skating is thrown into the mix. One common one is where a figure skater with their arms out is spinning on the spot (technically they're skating backwards in very small circles), and then bring their arms in to their body. This reduces the distance between the axis of rotation and some of the skater's mass, and due to the conservation of angular momentum, the rotational speed of the skater increases to compensate, that is, they spin noticeably faster. Another favourite is when talking about Newton's third law, that for every action, there is an equal and opposite reaction. This can be highlighted fairly easily in scenarios with less friction, so ice skating is again a prime candidate, perhaps with a scenario of an ice skater attempting to bowl or the like. Ice skating doesn't only enter the scene for the purpose of examples though, a fair bit of study has gone into the physics of ice skating itself.

So there's slippery ice and people on blades with a small surface area sliding across it, what's there really to study? Well actually, explaining the reason behind the thin layer of liquid water between the blades and the ice, which is what lowers the friction so much, turns out to be rather tricky. Applying pressure to ice lowers its freezing point, a phenomenon known as regelation, so an early idea was that the weight of the person on the blades applying a lot of pressure to a small area of ice caused a tiny amount to melt, but apart from not dealing with scenarios where the temperatures were sub-zero (Celsius), this still wouldn't be enough to explain the observed lowering of friction. When the numbers are all crunched, it turns out that it would take a skater weighing around 750 kilograms for this explanation to be enough on its own, something else must be at play. Suggestions have been put forward involving possible irregularities in the ice amplifying the regelation effect (the ice isn't atomically smooth after all), water molecules at the surface not being held as tightly as those in the rest of the ice (essentially vibrating more strongly due to having less neighbours) meaning it takes less to turn them into liquid, the friction from movement adding some additional heat to the equation, and multiple others. To this day though, whether one or a combination of these, or perhaps something we've not yet thought of is at play, the physics behind that thin film of liquid between blade and ice remains something of a puzzle.

## Fire

It's perfectly possible for an explanation to be both technically accurate, and not illuminating in the least. The now classic example is of one child[5] asking, 'what is a flame?' He was so disappointed with the initial answer of 'it's oxidation' that more than half a century later in 2012, he established an international competition, the 'flame challenge', to find a more satisfactory answer, one that would both give a clear understanding, and that an 11-year-old would easily

---

[5]Alan Alda

understand. The now yearly competition asks a different question each year, but to return to the original question, what is a flame?

If you light a candle, the carbon and hydrogen of the wick are given enough energy to turn into a gas and mix with the oxygen in the air. When they mix, this highly energetic process is called oxidation. Close to the wick, there's enough oxygen to go around for this process and the result is the atoms radiate blue light. Further from the wick, since some of the oxygen has been spent, there's less left to mix with the wick turned gas, the fire's fuel, and the result is leftover carbon. This left-over carbon makes soot, and when soot's heated, it radiates its excess energy as light. The more heat, the brighter the soot glows, which is what you're looking at in the red/orange/yellow part of the flame, huge numbers of tiny soot particles all radiating their excess energy as light. As they're more energetic than the cooler air around them, there's a buoyant force on the soot particles which is greater than their weight, so they rise, helping make way for more oxygen to feed the reaction and producing the familiar flame shape. If you were in zero gravity though, hot air weighs the same as cold air, so the soot particles would then feel no buoyant force on them. They wouldn't rise, limiting the amount of oxygen able to reach and feed the bottom of the flame, resulting in a much more spherical flame shape.

Fire needs three things, heat, fuel and oxygen. Take away any of these three ingredients and the fire stops. Counter-intuitively, it's also possible to extinguish a fire with a strong enough electric field, as the soot particles in the flame are easily charged, becoming ions which are affected by an electric field. To extinguish a normal fire, water is fine as it removes heat from the surface. With enough water it also submerges the fuel, depriving it of its access to oxygen. However, if added to a grease fire, since oil and water don't mix, the water sinks to the bottom and evaporates immediately spreading the flaming oil everywhere, leaving the removal of access to oxygen as the preferred option. Some fires are particularly tricky. If for example you were to try and put out a magnesium fire with water, the magnesium would react with the water to produce hydrogen, and the hydrogen would then ignite, essentially adding more fuel to the fire.

Historically, the trickiest fires to extinguish have been natural resource fires, with multiple coal ones considered near impossible to extinguish. The coal seam of Brennender Berg in Germany for instance has been burning continuously since the 17th century. With oil well fires, a once common practice for extinguishing them which can still be found in use today uses explosives like dynamite to force the burning fuel and oxygen away from the fuel source before then attempting the dangerous process of capping the well. However, the most extreme approach to extinguishing fire comes from the Soviet Union in the '60s where they once used a specially made 30 kiloton nuclear bomb to stop the burning of five natural gas wells.

## Water

If someone ever makes a surprising claim pointing to some unexpected phenomenon, a situation which has certainly not been uncommon throughout the history of science, the response preferred from a physicist, or any scientist for that matter wishing to investigate the claim is going to be the same, perform a repeatable experiment which confirms or denies said claim. The easier it is to set up and perform the experiment, the easier it is to confirm or reject what's being claimed. When it comes to water, due to its ready availability and ease of handling, experiments tend to be particularly easy to set up. For example, did you know that due to the viscosity difference between hot and cold water, you can actually 'hear' the difference between the two being poured into a cup? A lifetime of listening to liquids poured has allowed your subconscious to pick up on the difference, an easy experiment to design and try with a friend with a result that many find surprising. There does seem to be one claim that's an exception to all this however, that hot water freezes faster than cold water.

Water ($H_2O$) has some well-established and interesting properties. It's a liquid at room temperature, while other similar substances such as hydrogen sulphide ($H_2S$) remain gaseous. It also expands getting less dense when it freezes, allowing lakes to freeze from the top down and ice to float to the surface of water. So with regards to the possibility that hot water freezes faster than cold, it wasn't a huge stretch to imagine that there might be yet another thing to add to water's list of quirks, but when it came time to put it to the test, the experiment failed, hot water didn't appear to freeze faster than cold after all.

Normally that would be the end of the story and the claim could be dismissed, but over the years, reports of the phenomena kept coming in from a wide variety of sources. Different people seemed to be getting markedly different results, something made all the more confusing by the simple nature of an experiment with so few variables to control. While there have been many theories put forward over the years to try and explain why some experiments repeatedly seem to confirm while others deny the claim, to this day it still remains something of an unsolved mystery.

# 11. Earth's oceans

## The Great Pacific garbage patch

Garbage is a part of life for modern humanity. We create rubbish tips to put it in, and come up with ways to deal with as much of it as we can, from composting and recycling, to Sweden's approach of burning it as fuel for energy plants. Not all garbage is visible though. Plastics don't really break down over time, they can however break apart into smaller and smaller pieces, eventually becoming microscopic. It's not as simple as out of sight, out of mind though, and perhaps the best example of this is in the ocean.

Due to a combination of factors, Earth's Oceans have currents which cause the water to flow in predictable ways. A large system of circulating ocean currents is called a gyre, of which Earth has five major ones. As these gyres circulate, they pick up garbage in their path. Wind driven surface currents then gradually move this collected garbage towards the centre of the gyre, where it stays. If the garbage is plastic, it breaks apart over time, eventually becoming microscopic particles suspended in the water. The largest such collection is a combination of the Western and Eastern Pacific garbage patches, making what's known as the Great Pacific garbage patch. While this name can be a little misleading, as most of the 'garbage' isn't actually visible, it's the name which seems to have stuck. There are regions where visible floating rubbish does gather, extending for large distances, but the bulk of the patch is more of a 'plastic soup'. Due to this, the exact size of the patch is quite difficult to calculate as it can only be determined by sampling. Estimates, which are normally accompanied by a visual comparison, range anywhere from an area the size of Texas, to an area twice the size of the continental United States.

The patch obviously has a huge impact on the environment, killing or effecting a staggering amount of wildlife each year, and unsurprisingly this has a negative impact on humanity. As one of the more direct examples, some plastics can act as chemical sponges, attracting things like the pesticide DDT (dichlorodiphenyltrichloroethane), before entering the food chain, moving from toxic plastic, to jellyfish, to fish, to dinner plate. It can be a surprisingly simple process from first entering the ocean to coming back in your food.

## Waterfalls

What's the biggest waterfall on Earth? You may think that answering this would require specifying some more details. Does biggest refer to volume of water per second? Perhaps the question's asking about the distance the water falls, or the width of the waterfall? Actually for all these, the answer is still the same, the Denmark Strait cataract. So if this waterfall is really so big, why isn't it more well-known? Why have you never heard of it as a tourist destination before? It may at first sound counter-intuitive, but it's because the Denmark Strait cataract is an underwater waterfall.

So how can water 'fall' underwater? This comes down to temperature. Cold water is denser than warm water. The Arctic water on the eastern side of the Denmark Strait is colder than water on the western side, so at the ridge where the two meet the colder, denser, Arctic water flows downward, sinking underneath the less dense, warmer water from the western side. This type of downward water flow underwater isn't unique to the Denmark Strait cataract, but it is the largest known example of the process.

## Brinicles

If you've ever lived in a particularly cold climate, you're probably quite familiar with the idea of salting roads. The addition of salt lowers the freezing point of water, so adding it to roads limits falling snow or rain's ability to freeze into dangerous icy conditions. It's not impossible for ice to form from salty water though, as seen in the Arctic and Antarctic where salty sea water freezes into sea ice.

When ice is formed, it's always quite pure. As most impurities, including salt, tend to get expelled, polar sea ice is quite porous compared to regular ice. The expelled salt has to go somewhere though, and can leak into surrounding sea water, making it particularly salty. This further lowers the freezing point of the water, while also increasing its density. This supercooled, super salty water can then begin to sink into the ocean in a kind of plume, and as it does so, normal sea water coming into contact with it will freeze. This process can continue downwards creating a hollow, underwater icicle around the plume known as a brinicle (brine icicle), which in some cases can reach right down to the sea floor, an extremely dangerous prospect for any sea creatures like starfish unlucky enough to be in its path, as they're then frozen to death.

It's been proposed that perhaps brinicles could help life first appear in cold environments, in a similar way to the more commonly proposed method involving hydrothermal vents in hot environments. While still a relatively new idea, if true it would mean that some other places in the solar system such as Jupiter's moons Ganymede and Callisto would have much higher chances of having had conditions conducive to life at some point. Not guaranteed by any means, but still an interesting upping of their chances.

# 12. Astronomy

## Doppler shift

Trying to figure out exactly how far away something is in space, say a distant galaxy, was a tricky proposition for a long time. The technique currently used relies on a property of a special type of supernova known as type 1a, which involves two stars, one of which must be a white dwarf. This particular type of supernova will always reach the exact same level of brightness, or has a consistent peak luminosity, and so by examining how bright it appears from Earth, it's possible to tell how far the light has travelled, and therefore how far away they are. This in turn allows them to be used as 'standard candles', measuring sticks by which the distance of other things can then be judged.

Knowing the distance of something is only half the battle though. Astronomers also want to be able to tell if something is moving towards or away from us and how quickly. This is a much easier thing to tackle and relies on something known as the Doppler effect. You're probably familiar with this effect if you've ever heard an emergency vehicle's siren going past you on the street. The pitch seems to change, going from higher while it's approaching you to lower as it gets further away from you. Don't confuse a change in pitch with a change in volume, this isn't the same thing as when you're driving with the window open and hear the 'whoosh' sound while passing a parked car, that's just the sound of your engine being reflected back at you, with the volume going up when there's more reflected sound. The siren is continuously making the same sound, but while it's approaching you, the newer sounds reach your ears more quickly as there's less space to cover, the sound waves have bunched up, and the pitch goes up by an amount depending on how fast the emergency vehicle is approaching you. When it's receding away from you, the newer sounds have further to travel to reach you, and so the sound waves are more spread out, resulting in you hearing the siren at a lower pitch. The same thing happens with light, so when a distant star is moving towards us, its light waves bunch up, causing the light we see to be blue shifted or to have a higher frequency. Similarly, when the star is moving away from us, the light we see from it has been red shifted or has a lower frequency. Measuring the Doppler shift not only tells us if something is moving towards or away from us; but the amount the light has been Doppler shifted also tells us how fast this is happening.

## Supernovae

If you could take an outside view of any galaxy and fast-forward things, like watching a time-lapse video filmed through the millennia where every second is a 100 years or so, you would see the galaxy spinning, but it would also be twinkling. This is because in any given galaxy there is, on average, one supernova per century, and when there is, for a brief moment in time, one star can shine as brightly as the rest of the stars in a galaxy combined.

A supernova can occur when there is a change in the core of a star, which can happen via one of two understood mechanisms. The more common of these can be labelled either type Ib/c or type II. These are where towards the end of a star's life, as it runs out of fuel, some of its mass flows into its core. If the star is massive enough, this can cause a collapse when it becomes so heavy that it's unable withstand its own gravitational force. The other, rarer way is labelled type Ia and occurs when a white dwarf, a star that's out of fuel and would otherwise just gradually fade away, has a companion star. If close enough, as it orbits, the white dwarf can suck matter off its companion star until the white dwarf becomes too massive to support itself. When a star explodes in a supernova, it leaves behind a rapidly spinning, highly magnetized neutron star called a pulsar. There are a few different types of pulsar possible which astronomers categorize by the electromagnetic radiation they all emit via two jets being shot out in opposite directions.

Of the eight Milky Way supernovae in recorded history, one is particularly noteworthy. SN 1054, so named because it happened in 1054 AD, occurred some 6000 light years away. It was observed by multiple ancient cultures of the time, was as bright as the Moon, and would've been visible for a couple of years afterwards. The core of the star formed a pulsar called the Crab Pulsar while the resulting debris field is now known as the Crab Nebula. As these are both well studied by astronomers, being the two brightest objects of their kind visible from Earth, SN 1054 has become the most well-known, or even most famous of all recorded supernovae.

## Exoplanets

All through human history, we've been looking up at the stars. In modern times our equipment for viewing them has dramatically improved, but are there any real surprises left, or is technology merely allowing clearer images of what we already knew was there? As it turns out, the sky still has many surprises in store for us, as well as a few mysteries that we still haven't managed to figure out.

Probably the most famous space telescope is Hubble. In 1995, the then director of the Space Telescope Science Institute risked his position to focus it for a full ten days on a seemingly empty section of the sky no bigger than a grain of sand held at arm's length, allowing a particularly long exposure time. The end result was one of the most famous pictures in astronomy, the Hubble deep field (HDF). Far from seeing nothing, as was speculated to be a strong possibility by many at the time, there were over 3000 galaxies visible. This substantially changed our view of the universe, upping estimates of galaxies in the visible universe from around 10 billion to many trillions. Since then there have been two similar follow-up pictures focusing on a different patch of 'empty' sky, the Hubble ultra-deep field (HUDF) and the Hubble extreme deep field (XDF), with even more galaxies visible in an even smaller section of sky.

It's not just the number of stars that came as a shock in modern times though, the number of planets has also taken us by surprise. Planets orbiting stars other than our own, or exoplanets, generally can't be seen directly with current telescopes as they're typically around a billion times dimmer than the star they're orbiting. They do however block out a portion of the light of their star if they pass in front of it from our perspective, and this can be seen. If a distant star has a planet whose orbit takes it between said star and us, the star's brightness will dim by an amount proportional to the planet's size. This 'transit' as it's known will cause a dip of around 1% in the star's brightness for a Jupiter sized planet. Earth is around 11 times smaller than Jupiter and so would not cause much of a dip at all, but still enough to be measurable. Additionally during a transit, a sliver of the star's light passes through the planet's atmosphere, and with the right telescope, by determining which wavelengths are missing in the light spectrum we observe, it's sometimes even possible to tell what gases and chemicals are present in the planet's atmosphere. Before this transit method of detection was introduced in the '90s, we had no real way of accurately estimating the number of exoplanets out there. Now, having already found thousands, in fact in 2017 it was revealed that the star TRAPPIST-1 just under 40 light years or ~12 parsecs away has 7 of them, we can estimate that on average there's at least one planet around every star in the galaxy; a figure substantially higher than anything that would previously have been guessed.

Probably the most intriguing piece of data to come from the transit approach concerns what's sometimes referred to as Tabby's star, where three recorded light dips had three unusual properties causing astronomers to scratch their collective heads. First, a normal transit lasts a few hours, these lasted anywhere from a week with the first two to nearly 100 days for the third, which appeared to be more of a complex superposition of multiple dips. Second, they were asymmetric, not smooth as any spherical, or planet shaped body would produce. Lastly, the dips blocked anywhere up to a whopping 22% of the star's light, which would require something with an area over 1000 times that of the Earth. While theories from planetary debris fields to comets have been put forward to try and explain these dips, at the time of writing none have been accepted as entirely satisfactory, in fact it seems to be getting trickier to explain with data now suggesting that Tabby's star is also dimming as time goes on, leaving it to remain as quite the unexplained mystery for Astronomers.

## Starshades

Just because we can't see most exoplanets directly with current telescopes, doesn't mean it will always be the case, and the technology expected to address this in the not too distant future is the starshade. Normally when looking at a distant solar system, the light from the star of that system is blindingly bright compared to the very dim orbiting planets that we'd like to focus on. The idea then is to point a normal space telescope at the chosen solar system, but position an occulter a few tens of meters across, around 72,000 km or 45,000 miles in front of it. This would exactly block out the light from the star, hence the name starshade, but not the view of the rest of the solar system being studied. Orbiting planets which would normally be too dim to be visible could then be focused on.

In order to prevent the star's light from simply diffracting around the edge of the starshade rendering it useless, it has to be specifically shaped. The resulting design resembles something of a sunflower, with petal shapes around the edge. As far as space equipment goes, this flower shaped light blocker might not sound like the most high-tech gizmo ever, but if all goes to plan, it should reduce the light from stars being focused on by up to 10 billionfold, allowing us to directly observe exoplanets up to ~32 light years (~10 parsecs) away. This could possibly reveal details otherwise out of reach to us, such as the presence of oceans, continents or polar caps on these distant worlds.

## Gravitational waves

Until very recently, whatever approach you were using to spot something in the universe, whether looking at microwave radiation left over from the big bang, gamma ray emissions from supernovae, or just up at the Moon on a clear night, they all had one thing in common, a reliance on light waves of some frequency. This changed in 2016 when the advanced LIGO (Laser Interferometer Gravitational wave Observatory) announced the first detection of a gravitational wave.

When things move, they create waves. Shake your finger back and forth in the bath and you get water waves, shake (pluck) a string on a guitar and you get sound waves, shake some electrons back and forth and you can get radio-waves. If you could shake the Sun back and forth, you would get gravitational waves, but these, instead of travelling through water, air or the electromagnetic field, would travel through the fabric of space.

The first gravitational wave detected came from the collision of two black holes which spiralled around each other before merging around 1.3 billion years ago. When they collided they released ~3 solar masses worth of energy in a few milliseconds. To try to put this into some kind of perspective; in that moment, the energy released was equal to ~50 times of all of the light being emitted from all of the stars in the visible universe combined. Being black holes, this energy wasn't released as light or mass, it was instead pumped into the very fabric of space causing waves. So what happens when these waves pass through something? Well, ever see a cartoon character getting stretched in different directions, going from tall and thin to short and wide before going back to normal? This is actually a fairly accurate picture of what happens except for the scale. The fabric of space is quite stiff, extremely resistant to being stretched. By the time these waves reached the Earth, the change that needed to be measured to detect them was one part in $10^{21}$. If we were talking about the distance from the Earth to the Sun, this would be equivalent to measuring a change in distance about the size of a single atom.

The method used to measure this change relied on splitting light and bouncing it off mirrors in different directions ~4 kilometres or ~2 ½ miles away, recombining the beams once they returned and seeing exactly how the combined light waves matched up. Due to the incredible precision required though, it took an amazing feat of engineering. For example, most people know that the Moon affects the tides, but did you know that it also affects the land, moving the ground up and down around +/- 15 centimeters or 6 inches? As this tidal bulge passes by twice per day it stretches the Earth just a little bit, and if not accounted for, would change the distance the light had to travel to the mirrors by +/- 120 microns. This may sound like it would be insignificant, but even a change of this amount would have been enough to throw the entire experiment off and therefore had to be compensated for.

The frequencies of the gravitational waves we're looking for happen to lie in the same frequencies as those of sound waves that our ears are sensitive to. The type of wave is completely different, as are the mediums they travel through, but it's still possible to take the data and translate it into a sound that we can listen to. The sound of the two black holes colliding created what's famously known as a 'chirp'. Physicists like to extend the hearing analogy further. There were two LIGO detectors, one at Hanford Washington, and one at Livingstone Louisiana, and the difference in timing when recording the chirp allowed us to roughly figure out which direction it came from, much like listening with two ears allows you to get an idea of which direction a sound comes from.

So now that we've successfully listened to the universe, is that the end of the story? Has it come time to pat ourselves on the back and move on to other things? Actually, this is likely just the beginning of a whole new age for astronomy. Besides collecting more data on black hole collisions, there are many other events that we'd also like to study which have previously been impervious to examination due to our reliance on light waves; neutron stars whether spinning super-fast or colliding with other neutron stars, the centres of supernovae and even the early moments of the universe after the big bang. Plus, you never know what new and unexpected things you might discover with access to a whole new way of studying things. To say that some physicists are excited about the possibilities that gravitational wave detection has opened up would be a major understatement.

# 13. Earth's moon

## Synestia

For a long time, the leading theory for the formation of Earth's moon was the 'giant impact theory', where a Mars sized body hit the Earth in its early years, and the moon formed from the orbiting debris. The main problem with this theory was in the isotopes of the elements of the moon. They don't seem to come from some outside source, but instead, match the Earth exactly. The Earth and moon are made from identical stuff.

The theory as it stands now still involves an impact, but instead of picturing the result as the Earth surrounded by a debris field, imagine the Earth, as well as the body that hit it, completely vaporised. A super hot gas, spinning extremely fast, spreading into a slight disc shape. This is a synestia. It doesn't last very long, cooling back and shrinking into a planet in relatively short order, perhaps in just a couple of hundred years, but that could be long enough for the moon to have formed 'inside' the Earth as a synestia, condensing out of rock vapour, orbiting inside for a while, only to be revealed when the synestia cooled and shrunk back to form the Earth.

## Tidal locking

In the time it takes the Moon to rotate around its axis once, it rotates around the Earth exactly once. This means that no matter which night you choose to look up at it, the same side will always be facing the Earth. Our moon isn't unique in this regard though, given enough time, this situation, known as tidal locking, will happen to any moon around any planet. Pluto and its largest moon Charon are even tidally locked to each other, only ever showing one side to the other, but what causes this effect?

Recall that as the Moon passes overhead its gravity not only effects the tides, but the land as well. The Earth has the same effect on the Moon, but to a far larger degree, pulling on its otherwise roughly spherical shape, stretching it into more of a oblong shape with two bulges pointing towards and away from us. Originally the Moon was rotating much faster. As it rotated, the Earth's gravity was constantly attempting to reshape the Moon so that the bulges were facing towards and away from us, but reshaping vast amounts of the Moon's rock isn't instantaneous, it takes time, and so as the Moon spun, the bulges were always a little out of alignment with the pull of gravity of the Earth. You could imagine the extra mass of these off-centre bulges almost like handles. As the Earth's gravity grabbed on to one to pull it back into alignment, it slowed down the Moons rotation slightly, until eventually it succeeded, the bulges were aligned, and the Moon was tidally locked.

So if tidal locking happens due to the effect of gravity on the bulging of a celestial bodies' mass, and the Moon effects the Earth's land mass as it passes over head, does that mean it's just a matter of time until the Earth is tidally locked to the Moon in the same way that it is to us? Actually, the answer is that given long enough, yes, this would be the case, we would only ever show one side of the Earth to the Moon, so the Moon would appear to stay stationary in a single spot in the Earth's sky. Due to the difference in mass though, it would take an extremely long time for this to happen (~50 billion years). However, it does mean that Earth's rotation speed has been slowing steadily as a result of the gravitational effects of the Moon (and also the Sun). Days are getting longer by ~15 microseconds every year, and while this may not sound like much, over the course of Earth's history this effect has added up substantially. So much so that in the distant past, days on Earth would have only been around 6 hours long.

## Earth's 'second moon'

Ever heard someone talking about Earth's second moon? Cruithne (from Old Irish, pronounced KROOee-nyuh), is a near Earth asteroid first discovered in '86. It had remained undiscovered before this due to its small size, around 5 kilometres, or ~3 miles in diameter, making it dimmer than Pluto when viewed from the Earth. It orbits the Sun in an elliptical orbit which takes it inside Mercury's orbit and outside of Mars'. Cruithne's speed varies at different points in its orbit around the Sun, but a single revolution takes ~364 days, meaning it and the Earth tend to follow each other around. Due to this, it has sometimes been referred to as 'Earth's second moon', this is quite a misleading title though as Cruithne is not a moon at all. It doesn't orbit the Earth or any other planetary body, a pre-requisite for any celestial body being a moon.

A better candidate for the title of Earth's second moon would be the asteroid 2016 HO3, which also remained undiscovered until quite recently due to its small size. It too orbits the Sun, so is more accurately a quasi-satellite of Earth and can't technically claim the title of moon either, but its orbit is such that it never strays too far from Earth, even circling us from our point of view. From time to time Earth has had various such quasi-satellites, but most don't stick around for long periods of time. Asteroid 2016 HO3 is a bit of an exception in that regard though as it seems quite stable, probably having been with us for around a century, and predicted to stick with us for centuries to come.

## Helium-3

If humanity were to become a truly space faring species, what energy sources would we ultimately end up using beyond the Earth? In the short term we could transport fuels up from Earth's surface, but that's not a likely long term solution. We could sometimes use solar power, but what if we wanted to build a base on the dark side of the Moon, what fuel options for power would we have available then? In the long run, the most likely answer is helium-3.

Space has regular helium (helium-4) in it. Sometimes an energetic cosmic ray might hit some of this helium-4 knocking a neutron off, resulting in helium-3. The Sun also belts out large amounts of helium-3. The Earth has a magnetic field and an atmosphere meaning the helium-3 doesn't reach us here, but the Moon has no such protection, and over its ~4 and a half billion-year history, helium-3 has been settling like dust in the lunar regolith, the Moon's equivalent of topsoil. In just the top few centimetres it's been estimated that there would be enough helium-3 to power the Earth for ~1000 years, with even larger deposits available further out in the solar system if we ever get to places like Jupiter.

Due to the limited availability of helium-3 on Earth, with supplies mainly coming from the decay of tritium in nuclear weapons, limited research has been put into how best to harness the large amounts of energy locked up within it. Should humanity ever find itself colonizing other bodies in the solar system though, due to the natural abundance of this clean (non-radioactive) potential fuel source, research into maximising its use would likely become a high priority.

# 14. Radioactivity

## Carbon dating

While many topics in physics can come with a sense of wonder at how the universe works behind the scenes, the topic of radioactivity can instead bring with it a certain level of fear. This needn't be the case though, for as with any other danger in the world, with a little knowledge things can quickly appear much less scary. Much like dealing with electricity, where you wouldn't play with a downed power line, but you would unscrew a light bulb, it's just a matter of understanding where the danger is. It may seem easier to simply say that you'll play it safe and just avoid dealing with anything radioactive altogether, but this might be trickier than you think. Not only would this involve avoiding things like smoke detectors or exit signs, but you are actually radioactive right now. In the atmosphere, a radioactive form of carbon can be found, carbon 14. When its nucleus explodes it emits beta radiation, but more is constantly being generated by cosmic rays and this balances out so that carbon 14 remains at the relatively steady level of around 1 part per trillion. Plants take in their carbon from the atmosphere, animals eat plants, and you, whether you eat these animals or the plants directly, continually replenish the level of this radioactive carbon in your body with every meal. This then is how carbon dating a body works. It essentially measures at what level the carbon 14 is at in a body, and from that determines how long ago it stopped eating.

## Radiation

Inside the atom of a heavier element, there can be many protons all with a positive charge trying to push away from the other protons, and radioactivity is when the nucleus of such an atom spontaneously explodes in an extremely energetic reaction, usually with a million or more times the energy that any chemical reaction can have. When this occurs naturally we call the substance radioactive. It shouldn't be thought of as some timer counting down though. It comes from the probabilistic nature of quantum mechanics, so if a block of some substance has a half-life of 1000 years, on average after 1000 years you will be left with half as much of your block, but that half that you have left will be 'exactly' the same as it was when you started out. A nucleus is not more likely to explode if it lasts a long time, it has the same chance, second to second, day to day, remaining exactly the same as time goes on.

What comes flying out of these explosions is what we call radiation, and there are multiple different types which fall into two categories. The first type are charged particles and are ionizing radiation. They can be either alpha (α) particles, which are identical to the nucleus of a helium atom, with two protons and two neutrons, or beta (β) particles (or rays), which are electrons. The second type is photons of light at various energies, covering everything from radio-waves to X-rays and gamma (γ) rays. There are other types of radiation including neutron radiation, which as the name suggests are neutrons, and fission fragments, leftovers from nuclear fission. The sun also emits the solar wind which is made up primarily of free protons, but unless you're dealing with a nuclear power plant or bomb, or planning on leaving the Earth, you probably won't be encountering these types in your lifetime.

## Radiation damage

So now that the different types of radiation have been established, the next thing that people tend to want to know is how they might affect you. Well, if either an alpha or beta particle passes by another atom, because they're charged particles, electrons will either be attracted or repelled by this charge, potentially knocking the electron off an atom as they pass by and thus ionizing the atom. Some atoms behave very differently when they become ions. A water molecule for example might fly apart, but of course the real damage can come if the ionized atom is part of your DNA as it could break it. The cell may just die and not function properly any more, or worse it may mutate, possibly in just such a way that what normally inhibits its replication is turned off, resulting in cancer. Alpha or beta radiation from most sources you'll ever encounter isn't really something to worry about though, they just don't have the penetrating power to get to where they can cause any real damage, probably being stopped by just the dead layer of skin on the outside of your body. Photons of light can be another story though as they can be extremely high energy and thus potentially penetrate to where some real damage can be done.

Photons of different energies have various penetrating capabilities. Visible light for instance is stopped by your skin, whereas X rays penetrate your skin, but are stopped by the calcium in your bones.

If a high-energy photon of light collides with an electron, this colliding radiation can impart its energy to the electron, knocking it off and ionizing the atom, but worse, that freed electron is now itself ionizing radiation, a beta particle, but potentially starting deep within your body where it's able to leave a trail of ionized atoms in its wake.

When first encountering all this it may seem a little disconcerting, but actually, this may help many feel more at ease with things which they previously viewed as dangerous. Clearly not all photons of light are the same, and the type that probably gets the most undeserved reputation are microwaves. Microwaves are less energetic than visible light, which already doesn't have enough energy to knock an electron off an atom. Like any light, microwaves can still impart their energy to what they encounter and so can still warm things up, like in a microwave oven, but being unable to cause ionizing radiation there's absolutely no known mechanism where they can lead to cancer, no matter what media scare campaign you may have encountered in the past. In the same way you'd feel safe from any increased cancer chance from playing with a light bulb, you can now feel equally safe playing with say your mobile phone.

## The linear hypothesis

Once people know the types of radiation and how they can cause damage, it's natural for them to want to know the amounts needed to cause damage. You certainly don't need every cell in your body, but lose enough and you'll die from radiation poisoning, so how much is dangerous? For this some units measuring harm done to cells need to be introduced. 2 billion gamma rays per square centimeter = 1 rem, which as you can probably guess is a rather large dose to start the scale at. 100 rem = 1 sievert (Sv) and at this level you'll start to see some signs of ill effects or radiation sickness beginning to show. 3 sieverts to every cell in your body is called a whole-body dose which is a lethal dose 50% of the time within a few weeks of receiving it, known as LD50. At 10 sieverts a recipient will be dead within a half hour. Ok, so that's the upper end of the scale right, but how much radiation before you start seeing something like an additional cancer appear on the scene? The answer, which will take some explanation, is that on average, 25 sieverts = one cancer.

If that wasn't a misprint, then what's happening? How is it ever possible for someone to get cancer from radiation if it takes 25 sieverts, when they'll be dead within half an hour at 10 sieverts? The answer lies in the distribution of the 25 sieverts. If for instance you spread that 25 sieverts of radiation out over 10 people, they'll each receive 2.5 sieverts worth, and each person will have an additional 10% chance of developing cancer on top of any normal chance they might have, so on average you will see one additional cancer appearing amongst the group. If you were to spread the 25 sieverts out even further between 100 people, so they each receive 0.25 sieverts or 25 rem, then each person will have a 1% additional cancer chance than they had previously, and again, the 25 sieverts will still result on average in 1 additional cancer appearing. This linear correlation continues no matter how many people the 25 sieverts are spread between and is known as the linear hypothesis. While it does occasionally get debated when talking about extremely low radiation levels, it being a theory which ethically is not possible to put through rigorous testing, it is generally seen as accurate and thus used as a crucial tool when making policies or assessing radiation risks.

## Backscatter X-rays

Ever read a comic or watch a movie where the superhero has X-ray vision? Ever wonder how that would work? If you've ever had an X-ray taken at the dentist's, you've probably noticed that there needs to be a plate behind what's being X-rayed. The X-rays are emitted, pass through your skin/gums but hit the calcium in your teeth, so the plate on the other side has an outline left behind. The superhero doesn't have a plate behind what they're looking at to work with though. Ok, so it's a fictional character, and the writers probably didn't put much thought into the scientific workings of the super-power, but still, there is a way it could possibly work, if their X-ray vision wasn't using normal X-rays, but backscatter X-rays.

As the name implies, backscatter X-rays collect the radiation that's reflected off certain materials. These form a 2D image, allowing a device to scan something while only being on one side of a subject. Due to X-rays being ionizing radiation, this has led to large debates over their use in airports. Are they dangerous? This is a good example of where the linear hypothesis is debated at low doses. There are those that argue for and against, with good points made on both sides. As for their use, while it can still depend on which country you're visiting, many large countries like America, and most recently China, have followed the European Union's move of banning them at airports, with the competing millimeter wave scanners which don't use ionizing radiation now being the preferred option.

# 15. Protection

## Radiation tablets

The final thing people tend to want to know about when it comes to radiation is defence. After the Fukushima disaster, you may have heard of shipments of tens of thousands of 'radiation pills' being shipped to Japan, but what exactly are they, and how do they work?

The human thyroid, found just below the Adam's apple, concentrates iodine. This is a problem if that iodine is radioactive, as seen by a marked increase in thyroid cancers years after the Chernobyl disaster in the areas most heavily affected by the accident. People under 40 are particularly at risk from this due to the way the thyroid works. Radiation pills are just potassium iodide, the idea being to flood the system so that the thyroid is fully stocked up on stable iodine for a couple of days, allowing any that's radioactive that someone may ingest or inhale during that time period to pass through their system instead. These pills don't lower the risk of other cancer types, only radiation induced thyroid cancer, and unlike some movies may have you believe, are purely preventative in nature. Unfortunately, while some promising progress is being made in the area, there's currently no known way to reverse cell damage already received from radiation. There may be one day, but for now, the protection offered by the radiation pill remains the best response option that we have available for people in high risk areas after a nuclear accident.

## Body armour

From the heavy suits of armour of medieval knights designed to deflect the piercing or slashing from arrows and blades, to the lighter ballistic vests of more modern times designed to deform incoming bullets, spreading their force over a larger area, the development of personal body armour throughout human history is possibly best summed up by the old phrase 'necessity is the mother of invention'. Unsurprisingly, the choice of materials has changed to keep up with the times over the years. In recent history, Kevlar dominated the scene due to its high strength to weight ratio. A century or two ago though it was silk, often worn into duels for its durability, minimizing the chance of clothing tearing into shreds and getting stuck in the wound when shot or stabbed. There are stories about Mongolian horse archers also wearing silk shirts under their attire which would wrap around incoming arrows without tearing, allowing them to tease the arrow out of the wound the way it went in by tugging on the shirt. While a possibility against rounded tip arrows which wouldn't puncture the silk, that's not the type they would have been encountering, suggesting this is most likely inaccurate and just an urban legend. Looking ahead though, what's next? Can we guess at the next big thing that will likely dominate when it comes to personal body armour? Unlikely as it sounds, it probably won't be a material at all, but a liquid.

Non-Newtonian fluids don't follow the normal rules for liquids, or more specifically, don't follow Newton's law of viscosity, they can change their viscosity when agitated. You can think of a fluid's viscosity as a measure of its thickness. Water is not very thick and so has quite a low viscosity, while honey is comparatively quite thick and thus has a much higher viscosity. A common example of a non-Newtonian fluid would be tomato sauce or ketchup. You may have trouble getting it out of the container initially, but after some shaking or tapping, it just flows everywhere, it's viscosity has decreased. If a non-Newtonian fluid's viscosity doesn't decrease, but rather increases in response to agitation, known as shear strain, it's a special type of non-Newtonian fluid known as a shear thickening fluid (STF), and it's this type of substance which will most likely be key to the next generation of body armour.

You might be familiar with the idea of a shear thickening fluid if you've ever played with oobleck before, made by mixing 2 parts water with ~3 parts corn starch. You can slowly dip your hand into it and it will be a liquid, but if you suddenly squeeze or punch it, it will stiffen, feeling solid. This is because of the suspended particles in the fluid. When there's no stress, they're free to flow along without any problems, like logs slowly floating down a river. When agitated though, the speed of the river increases, the logs bunch up and none can get through, you get a log jam. Specifically designed shear thickening fluids can have such extreme log jam type reactions to agitation that a single centimetre can stop a bullet, and there are a couple of different methods developed so far to capitalize on this feature. One is to soak a fabric in the STF, which results in a cloth virtually impervious to being punctured from stabbing or the like. Kevlar is then the prime fabric candidate to be soaked, leading to more effective body armour which can be designed with less material, in turn allowing more manoeuvrability for the wearer. Another approach is to have an entire layer, perhaps in pockets, of

104

STF inside the body armour. This way when hit by a bullet, not only is it completely stopped, negating the chance of any potentially lethal ricochets, but its impact energy is dispersed, rippling out into the entire rest of the armour's liquid. Spreading the shockwave out over such a large area in this way means the wearer won't be coming away with the usual nasty bruise, feeling like they were just hit by a hammer. Whatever the approach though, it looks fairly certain that shear thickening fluids are the next big thing in the field, and it's a fair bet that any top end body armour in development for the foreseeable future will be incorporating them in some form or another.

## Faraday cages

Is the inside of a car a safe place to be during a lightning storm? Assuming it's not a convertible or the like, and the doors are closed so that when inside you're mostly enclosed on all sides by metal, then it's probably one of the safer spots to be. This is because the car is essentially acting as a mobile Faraday cage. Don't go out driving and hunting for lightning storms just yet though, a car isn't designed as a proper Faraday cage, but if the storm is upon you and the choices are either stand around outside or seek shelter somewhere, you could do a lot worse than climbing into your car, turning everything off and waiting it out.

A Faraday cage works by acting as a hollow conductor, with the charge staying on the outside surface. Imagine a Faraday cage surrounded by an external electric field. The electrons in the metal of the Faraday cage get pushed away from the negative part of the charge to the far side of the cage, but a clumping of electrons in one area itself makes for a negative charge, and where they were pushed away from, the metal now has less electrons and so is positively charged. This means that the cage is responding by creating its own electric field. The stronger the outside charge, the more the electrons are pushed to one side of the cage, and the stronger the charge created by the cage, with the result being that inside the Faraday cage, the external electric field and the one created by the cage always balance out to zero. The most common use of a Faraday cage in everyday life is probably as part of a microwave oven, which reverses things somewhat, protecting the outside from the microwaves within. If you look closely at your microwave's door, you'll be able to see that inside the window there's a fine mesh. This is part of a Faraday cage which is completed when the door closes.

While a Faraday cage offers near perfect protection from an electrostatic field, static as in unchanging, like that from a lightning bolt, they're not so crash hot when dealing with electromagnetic fields, or fast changing electric fields, like that from your cell phone. The protection offered against electromagnetic fields is heavily influenced by two factors, the size of the holes, if any, which relates to the size of the wavelengths which can penetrate, and something called the skin effect. Essentially, electromagnetic (EM) waves work their way from the outside of a metal in, and different metals require different levels of thickness to handle different strength waves, with protection against lower frequencies generally requiring thicker shielding.

# 16. Categories

## Nuclear weapons

When talking about nuclear weapons, whether in the cinema, the media, or otherwise, words like A-bomb, H-bomb, nuclear bomb and thermonuclear bomb often get muddled up, being used interchangeably when in fact they can refer to specific things. Yes, a bomb is still a bomb, but there are some major differences which can mean the difference between an explosion equal to ~1 ton of TNT, compared to the record held by the 'Tsar bomber', a 1961 USSR test with an explosion equivalent to over 50 megatons of TNT (1 megaton = 1 million tons).

While there are some variations, nuclear bombs fall into two basic categories. The first combines two pieces of enriched uranium, sometimes known as the gun method as one piece is essentially shot into the other. When combined, they reach a mass large enough to start a nuclear chain reaction and fission ensues. This method is generally seen as easy to design and build, but difficult to acquire the uranium for as uranium enrichment is not an easy process. The second uses an implosion method where a mass of fissile material, generally plutonium, is compressed with explosives evenly from all sides until fission begins. This method is almost the opposite of the first in that it is extremely difficult to design and build, but with the required materials, plutonium, being relatively easy to acquire, at least when compared to enriched uranium. Both these bomb types fall under the category of atomic bombs (A-bombs).

Atomic bombs start to reach their explosive limit at the equivalent of around 500 kilotons of TNT (1 kiloton = 1 thousand tons). In order to get the big explosions measured in megatons, a second stage in the explosion process is required, the introduction of hydrogen isotopes, hence being informally called hydrogen bombs (H-bombs). More technically, they're thermonuclear weapons, as when the original fission occurs, it causes the hydrogen to undergo fusion at thermonuclear temperatures, releasing an enormous number of high speed neutrons.

Theoretically, any number of additional stages can be added. More hydrogen undergoing fusion means a bigger explosion, but realistically, things like the weapon's weight or missile warhead space add practical limits to design. Also, if it's not going to be used any time soon, there may be the fact that nuclear weapons have a shelf life to consider. They all rely on fissionable, radioactive materials, and that material's half-life means that given long enough, there won't be enough material left to go critical, it will simply cease to be a bomb.

## Star types

When talking about different stars, a lot of terms can be thrown around without stopping to talk about what they actually mean. Sometimes you can guess from the context, a red giant sounds large, while a yellow dwarf sounds comparatively small, but how many different types of star are there? In the current system, seven different types of star are recognized, determined by their temperature. These are assigned letters which, ranging from hottest to coldest, are O, B, A, F, G, K and M. This lettering system is left over from a previous way of categorizing stars according to their hydrogen content, managing to stick until the present day despite now referring to an unused system and probably being more confusing than useful. Additionally, stars are then designated a number within their category ranging from 0 to 9 (with fractional numbers allowed), with a lower number being hotter, so an A-9 star is cooler than an A-8 star, but still hotter than an F-0 star. The category given to a star does roughly equate to a particular colour which dominates the light spectrum of the star, however using just the simplified colour system can be misleading. A star radiates all colours of the light spectrum, which to the human eye is generally going to appear as white or perhaps slightly blue. A red star can still appear redder than a blue star, but not by the margin that colour designations alone may indicate. As a guide though, type-A stars can be set as 'white', leaving an O-type star as comparatively blue, B as blue/white, F as yellow/white, G as yellow, K as orange, and M as red.

So that's the temperatures/colours of stars covered, but what about star size? This is addressed with a different, more intuitive category system. Technically it categorizes a star's luminosity, but with two stars of equal temperature, a more luminous star equates to a larger star. The possibilities then, ranging from largest to smallest, are Ia, Ib, II, III, IV, V, VI and VII. Translating these to the more colloquially familiar, types Ia and Ib are supergiants, types II and III are giants, type IV is a subgiant, type V is a main sequence, or dwarf star, type VI is a subdwarf, and type VII is a white dwarf. These are not completely hard and fast rules, there is a small amount of flexibility when using this system. For example, you might occasionally hear of a type 0 hypergiant, or a type Ia/b star. Sometimes a type VI will instead be designated by the prefix 'sd', or a type VII with the prefix D, as white dwarfs don't need to be categorised by temperature the same way that other stars do. There are also many more possibilities for detailing information about a star which can be used, with white dwarfs alone having more than a half-dozen subcategories to indicate the composition of their outermost layer. Generally speaking though, using a combination of these two category systems will be enough to instantly indicate what you're referring to for any astronomer.

With all this, most stars can be easily categorized. The Sun we see every day for example is a type G2V star, or more colloquially a yellow dwarf. Is it possible though that there are stars we're yet to discover that can't be categorized by this system? While it's always possible there's something we don't fully understand, leading to a surprising discovery just around the next corner, stars do have limits, they can't get arbitrarily big for example. While this means it's extremely unlikely that we'll encounter anything that can't be covered, there is, technically speaking, at least one type of star that can't be categorized by the system. On a universal time scale, our universe is still quite young, and not every type of star that will ever be has had time to come into existence yet. Our Sun for example will one day begin to die. As it does so, it will expand into a red giant before shedding its exterior once all its nuclear fuel is spent. Being a relatively small star in the scheme of things, it won't be able to create a black hole, it will instead share the same fate as ~98 percent of other stars and leave behind a white dwarf, a small, hot, dead star which will begin to slowly cool. Very, very slowly. It will take tens of hundreds of billions of years for the white dwarf to lose its heat, but eventually, in theory at least, it will turn cold, and what will be left behind is a black dwarf. Not something which can be categorized by our current system, but certainly not something which necessitates a change to it any time soon.

## Pluto

In 2006, our solar system went from having 9 planets to having 8. This wasn't the result of some catastrophic stellar event though, rather a controversial redefining of what 'planet' actually means, with the upshot being that Pluto was no longer considered part of the planet club. Why the change though? And is that the last word on the matter, or are things likely to change again down the track?

This isn't the first time something like this has happened, and likely won't be the last. In the mid-19th century, Ceres was reclassified from planet to asteroid due to the discovery of many more objects with similar orbits, all part of the now known asteroid belt between Mars and Jupiter. Similarly, when Pluto was first discovered in 1930, we had a lot less information about objects in the solar system than we do now, and as time has progressed, many new objects much like Pluto have been discovered in what is now called the Kuiper belt, a region beyond the orbit of Neptune that's much like a giant asteroid belt. In just the years between 2002 and 2005, Quaoar, Sedna and Eris were all announced, all with comparable masses to Pluto.

So if reclassification is a normal process as we discover more and more celestial objects, what makes the decision labelling Pluto a dwarf planet controversial? Under the new definition, to be a planet a celestial body has to meet three criteria. It has to be in orbit around a sun, roughly spherical due to its own gravity, and the one which relegates Pluto to the position of dwarf planet, it also has to have 'cleared the neighbourhood around its orbit'. While various concerns have been raised about these criteria, and it certainly doesn't help things that so few astronomers were present for the actual vote, it's this final stipulation requiring a planet to have vacuumed up or ejected other large objects in its vicinity of space which is so controversial. The most commonly cited concern probably being that as you go further out from the centre of a solar system, it takes more and more mass to become gravitationally dominant. This means that a planet ceases to be about composition and becomes more about location. If Earth were to be in orbit in the Kuiper belt for example, it too could be considered a dwarf planet. While it's unlikely that this stipulation will be simply removed as that would lead to some confusion with more than 50 objects in our solar system as planets, and potentially hundreds more yet to be discovered, given the debate, it wouldn't be surprising if alternate suggestions for the definition of a planet were voted on in the not too distant future. Don't go completely dismissing Pluto's planet status chances just yet!

# 17. Waves

## Sonic booms

What do a whip, a lightning bolt and a Concorde airliner have in common? They can all produce sonic booms. The cracking of the whip, the thunder following a lightning bolt and the boom of a Concorde flying overhead at full speed are all examples of this phenomenon, a result of any object creating shock waves as it travels through the air at supersonic speeds, that is, faster than the speed of sound (mach 1), around 343 metres per second in air (varying slightly based on things like the air's humidity and temperature).

If an object creating sound waves is travelling through the air faster than the speed of sound, it's catching up to the sound waves it's generating ahead of it. The waves are then getting bunched up. Unable to get out of each other's way, they get compressed together until they merge into a single shock wave. An interesting side effect of this is that for the pilot of a plane travelling faster than mach 1, it's quiet. The sonic boom is a result of the sound waves left behind by the plane, much like the wake of a ship. The waves can't catch up without what's generating them slowing down first.

While hearing a sonic boom from a plane nowadays is extremely unlikely due to regulations ensuring jets don't exceed the speed of sound (too loud and disturbing for people in the 'boom carpet', the area that can hear the sonic boom), if you ever find yourself in the vicinity of a space shuttle landing, listen out for a distinctive double boom. This is because a supersonic craft forms two shock wave cones, one at the nose and one at the tail. For small craft these might be so close together that they're heard as a single boom, but for larger craft like a space shuttle, the first change in pressure boom from the nose reaching mach 1 and the second change in pressure boom as the tail of the craft passes and air pressure returns to normal should be easily distinguishable as two distinct booms.

## Cherenkov radiation

If you ever get to see the inside of a nuclear reactor that's working (perhaps by film or by photograph), then you'll likely notice that the water glows a striking blue. This distinctive colour is due to Cherenkov radiation, which could be seen as the light equivalent of a sonic boom.

Nothing can travel faster than c, the speed of light in a vacuum, but light slows down through different mediums, only travelling through water at around 3/4 c. Just because light's slowing down for a bit of a stroll though, doesn't mean the speed limit of c has changed for everything else, and in the water of a nuclear reactor, there are some extremely high energy, fast moving particles whizzing about, and some of these, while not breaking the speed limit of c, are travelling faster than the light is moving through the water.

The water of a nuclear reactor is there to absorb chargeless neutrons, but it's the beta particles (fast moving electrons emitted by fission products) moving through it which causes the glow. Beta particles are charged. When a charged particle moves, it takes its electric field with it, which is propagated by photons. If the particle is moving through a transparent, electrically polarizable medium (like water) faster than these photons can travel, then it's much the same as when a plane is moving through the air faster than the sound can travel, but instead of merging sound shock waves resulting in a boom, you get merging light shock waves resulting in a flash. A single particle gives a flash, the continuous glow in the reactor is due to the huge quantities of beta particles travelling through the water at any given moment.

Most Cherenkov radiation is invisible to humans. It covers a large range of light spectra, becoming more intense with higher frequencies. The human eye can't pick up on the intense ultraviolet portion of the radiation. We're also not very good at picking up on the violet end of the spectrum (at the edge of our colour range). It's not as intense at lower frequencies, so to our vision, blue dominates overall. This makes the blue glow somewhat unique. It's not like when you normally see something blue, with lightwaves corresponding to a particular spectral peak, the blue glow in the nuclear reactor has no single lightwave peak, it's continuous, and the resulting blue is just our vision's interpretation based on its intensity and what our eyes are most sensitive to.

## Sonoluminescence

A pistol shrimp (or snapping shrimp) has an unusual hunting technique. When it snaps its specialized claw at around 100 kilometres per hour (~62 miles per hour), it's so loud that a tiny cavitation bubble is created in the water. A cavitation bubble isn't the result of gas in the water like a normal underwater bubble, it's the result of a particularly rapid change in pressure. Think of an underwater sound wave as being a series of points with high and low pressure. With a loud enough sound then, a region can experience such intensely low pressure that the liquid is pulled apart into vapour. This cavitation bubble quickly collapses due to the pressure around it as the water rushes back in, with the resulting shock wave stunning or even killing the shrimp's prey, but what makes it so interesting from a physics perspective is a small amount of light that's released from the bubble when it collapses, referred to as sonoluminescence. The light isn't visible with the naked eye, but it's there.

Sonoluminescence can be easily reproduced in the lab. Some noble gas in the bubble also makes the light easily observable unaided. The bubble can also be kept at the desired point in a standing wave, causing it to collapse and reform repeatedly. With each collapse, light is generated, and the collapses are so frequent, several thousand times a second, that a seemingly continuous glow results. Ok, so we can easily maintain a bright blue light in a jar of water using nothing but a sound wave, but how exactly does it work? While many suggestions have been put forward over the years, it's yet to be shown conclusively exactly what's happening. Even the maximum temperature the bubble can reach is a matter of some debate. Measurements put the surface of the bubble around the same temperature as the surface of the Sun, but the temperature inside the bubble is a lot trickier to measure. Don't worry if you're ever near such an experiment though, while the bubble's temperature is extremely high when it collapses, the collapsed bubble is also extremely minute, around a single micrometre in diameter, so the energy released is still quite small in the scheme of things. No chance of lab accidents where the jar's liquid suddenly flash boils away.

Probably the simplest explanation proposed for sonoluminescence involves the collapsing bubble generating an imploding shock wave, compressing and heating the interior of the bubble to extremely high temperatures, hot enough to become a plasma. Proposed theories are quite wide ranging though, with one even suggesting that sonoluminescence is the result of virtual photons in the vacuum being converted to real photons due to the rapidly moving border between the bubble and the liquid, in much the same way that Hawking radiation is generated at a black hole's event horizon. While it will probably be some time until we know the exact mechanism with certainty, it's amusing to consider that for all we know, the pistol shrimp may actually have the rather impressive ability to convert virtual photons into real photons. Not a bad day's work for a humble crustacean.

## Sound channels

You may know that whale's songs are able to be heard across vast distances in the ocean, but have you ever wondered how they manage to achieve this? They're making use of something called a sound channel, which comes about from a combination of two factors. Firstly, a sound wave in water will bend, changing direction depending on the speed of the wave. More specifically, the wave will bend towards where it travels slowest. An analogy which is sometimes used for this is that of marching soldiers. If the soldiers on one side of a row are marching more slowly than soldiers on the other side, then the entire row will bend towards the slower side as they march forwards. Secondly, as you go down into the ocean, the water gets colder and comes under more pressure. Up near the surface this causes the speed of sound to be quite fast. As you go deeper, for the first kilometre or so, the speed slows down, but past this depth, if you continue to go deeper, the speed will increase again. Combine these two factors together and it means that for sounds made at around the depth of one kilometre, about half the sound gets trapped. If the sound has an upward component, taking it towards the surface, it can bend back down towards the region of slower speed. Likewise if the sound has a downwards component, taking it towards the ocean floor, it can bend back up, again towards the region of slower speed. Fluctuating like this back and forth, up and down, the result is that the sound stays at around the same depth, travelling within the sound channel, and can be heard over much greater distances than would otherwise be possible because the sound is essentially only spreading out in two dimensions instead of in three.

There's a similar sound channel in the atmosphere. In a gas, the speed of sound is slower than in a liquid or a solid. You can think of this as each individual molecule having further to travel in order to pass on the sound vibrations to its neighbours. In a warmer gas, the molecules have more energy, and so they can reach their neighbours sooner, and sound travels faster. Similarly if the gas is cooler, sound travels slower. The air just above the Earth is generally quite warm, cooling as you go upwards. This trend continues until you reach the stratosphere where the ozone layer is encountered. Here is where the Sun's ultraviolet rays are absorbed and so the air is again quite warm. Just like with the ocean, there's a sweet spot, an altitude where the sound travels slowest, resulting in a sound channel.

While the sound channel in the ocean, sometimes known as the deep sound channel (DSC), has been used for multiple things over the years, from signalling the need for rescue by downed pilots to submarine warfare, the sound channel in the atmosphere was only ever used briefly to listen out for nuclear bombs being tested. When a mushroom cloud from a nuclear explosion reaches the sound channel, the roar can be heard around the world. Technologies in other areas quickly superseded this method for detection and pinpointing the location of nuclear detonations though, and since then, the sound channel in the atmosphere has remained something of an unused curiosity.

# 18. Natural disasters

## Hypernovae

In the 1960s, the United States put Vela satellites into orbit. These were capable of detecting gamma radiation and were meant to tell if there were any secret nuclear weapons tests being conducted. They did detect gamma radiation, but to the surprise of everyone involved, it didn't seem to be coming from the Earth. After enough was detected that a direction could be established, it was determined that it wasn't even coming from the Milky Way. What event in deep space could generate such a powerful gamma ray burst though?

For around every hundred thousand supernovae, there is one hypernova, the result of a particularly massive star collapsing into a black hole. As you might expect, these are rare in the scheme of things, but the universe is big, and these events are detectable across vast distances, so on average Earth gets around 1 gamma ray burst per day from hypernovae. It's estimated though that for every one we see, there are several hundred more that we don't. This is because hypernovae, like supernovae, have two jets, where matter is shot out at high speed in opposite directions. The gamma ray bursts we can detect are also shot out these twin jets, so we can only detect them if one of the jets happens to be aimed straight at us.

While it can be difficult to grasp the scale of just how powerful these explosions are, consider this. If one of these hypernovae with its jets aimed straight at us were within even a few hundred lightyears of us, the gamma ray burst we'd experience would make extremely short work of all life on Earth. It's been seriously suggested that this could be the reason that Earth has yet to find signs of intelligent life elsewhere in the universe. Perhaps most life is wiped out by hypernovae before it gets a chance to evolve intelligence and develop technology. If that's the case, then Earth's been fortunate so far, and it looks like our luck is continuing to hold, for while astronomers are keenly watching a few stars in our corner of space which look to be nearing the end of their lifespan and could potentially produce a hypernova any time now, none seem to have their poles pointing our way.

## Volcanoes

Few natural occurrences can even compete with the powerful show nature puts on with an active volcano. Deep below the Earth's surface, the temperatures are so hot that rock melts into a mixture of gases and liquid (magma). This mixture is lighter than unmelted rock and so rises towards the surface, flowing upwards through a created network of interconnected chambers, attempting to push up through the weakest points in the Earth's crust. This is why a volcano can erupt multiple times over its lifespan, the path the magma rose through previously makes an easier pathway for new rising magma to take. When the pressure builds to sufficient levels, the magma bursts through, and the result is a volcano.

Volcanoes are categorized in various ways, from the composition of released lava to a system categorizing different stages of volcanic activity. They can also be categorized by their Volcanic Explosivity Index (VEI). If the magma is unable to break through the crust, the magma pool and associated pressures can just continue to grow. If this pressure builds long enough, it can result in what is sometimes referred to as a supervolcano, capable of truly world changing eruptions, having ejecta (particles like rocks that get ejected from the volcano during an eruption) with a mass greater than $10^{15}$ kg ($10^{12}$ t). One of the largest such eruptions occurred slightly under 28 million years ago in Colorado, U.S., with the total volume of ejecta equalling some 5 thousand cubic kilometres worth.

Supervolcanoes erupting on Earth are rare, the last one occurring in Indonesia some 74 thousand years ago. That being said, you may have heard that the Yellowstone supervolcano in Yellowstone National Park, possibly the most well-known of all supervolcanoes thanks to various movies over the years, is on a cycle and is now overdue, but don't worry too much, as 'overdue' when talking in geological timescales can have a somewhat different meaning compared to the usual use of the word. Its last eruption occurred roughly 640 thousand years ago, while it's cycle is expected to have an eruption once every 600 thousand to 700 thousand years, so yes, you could call it overdue, but thousands of years could pass without an eruption and Yellowstone would still be about as 'overdue' as it is now.

Other bodies in the solar system aside from Earth also have volcanoes, the largest of which so far discovered is on Mars. Named

Olympus Mons, it stands over 20 km tall (~2.5 to 3 times the size of Mount Everest) and covers an area around the size of Italy, but possibly more interesting than large scale volcanoes are the probable cryovolcanoes in the solar-system. While not yet directly observed, strong indirect evidence is mounting for their existence, the idea being that on a cold body like Ceres (currently considered a dwarf planet, much like Pluto), a volcano could erupt a liquid made of volatiles like water instead of silicates. The substance would be cold by our standards, hence a cryovolcano, but it would be hot compared to its surroundings in the same way that magma here might not be hot from the Sun's perspective, but is hot when compared to its surroundings on Earth.

## Tsunamis

The speed of sound waves is determined entirely by the medium they're travelling through. Their speed is completely independent of their frequency. If you're listening to music with simultaneous high and low notes, the music won't sound disjointed when you move further from the music's source. This is not the case for all waves though. If you've ever been at the beach and watched the little ripples on the water, their speed is substantially slower than the big waves rolling in. Different frequency waves in water can have different speeds. The depth of the water also has an effect, with shallower water decreasing a wave's speed. However, the wave is still carrying with it the same amount of energy, so when the wave passes an area with less water depth, energy is conserved by the wave increasing in height. This wave shoaling is what causes waves to break at the beach, as the now higher waves have tops which are travelling faster than the water below.

A tsunami, aside from being generated by water displacement due to an earthquake or the like instead of the usual wind or tidal processes, is unusual compared to other water waves due to their extremely long wavelength, sometimes up to 500 km between peaks. It's so far between peaks that even though they can travel at near the speed of sound, if you were out boating in the ocean and one went past, the water level would rise and fall so gradually that it would be near impossible to notice. However, once they reach land, it's a very different story. Wave shoaling also applies to tsunamis, resulting in an increased height of the wave as it approaches the shore, although unlike normal waves, tsunamis don't tend to break. While the hit of a tsunami wave can be damaging, sometimes causing secondary problems, the most severe example of which is probably Fukushima, the real danger generally comes from their long wavelength. Even at high speeds, it can take a long time for a full tsunami wave to arrive, and during that time, the water continues to rise, much like a rapidly rising tide (this similarity is behind them once being referred to as tidal waves). If the water level were to rise quickly but only for a few seconds, it would probably be manageable, but a tsunami will cause the water level to rise steadily for typically around 6 minutes which is another thing entirely. Finally, a tsunami also has multiple peaks to contend with which can hit the shore over several hours, and between these, when the water recedes, it will try and drag what's in it back out to sea. A particularly deadly combination of factors for anyone unlucky enough to be caught in a tsunami's path.

## Asteroids

Of all the possible natural disasters, the one probably most capable of catching us off our guard to cause the greatest loss of life on Earth is that of a large asteroid impacting the planet. The most well-known such event occurred nearly 66 million years ago when the dinosaurs were wiped out. The asteroid that hit then was about 10 km or 6 miles in diameter and impacted near what is now Chicxulub, Mexico, leaving a crater ~180 km in diameter, but was that just a one-off freak occurrence? In fact the Earth is hit by asteroids all the time. Around once a year a car sized asteroid will burn up in the atmosphere, but throughout Earth's history, big asteroids have not been uncommon, sometimes leaving substantially larger craters than the Chicxulub asteroid left. While not yet confirmed, the largest suspected impact crater is near Uluru/Ayres Rock in the Northern Territory, Australia. Named the Massive Australian Precambrian/Cambrian Impact Structure (MAPCIS), it would be 545 million years old, measuring ~600 km in diameter.

Asteroids come in all shapes and sizes, from just a metre or 2 to nearly a thousand kilometres across. One named Chariklo, in orbit around the Sun between Saturn and Uranus, even has its own rings. The majority of known asteroids are in the asteroid belt between the orbits of Mars and Jupiter, a region containing several million asteroids, with detection an ongoing process. Hopes are (or were, depending on what year you read this) set on ~90% of Near Earth Objects (NEOs) capable of causing a mass extinction being catalogued by 2020. An impressive feat, though unfortunately we'll never know the paths of 100% of all the asteroids out there. In 2017, the first ever interstellar object to enter our solar system was detected. While astronomers weren't completely surprised by this asteroid, named Oumuamua, and formally designated 1I/2017 U1, as they had been on the lookout for such an object for years, we still would have had very little warning if its path had been on a collision course with Earth. Generally speaking though, for non-interstellar objects, once catalogued, an asteroid's future path can be predicted quite accurately. The asteroid Apophis for example is one such tracked object. At ~325 metres in size, it caused some concern initially due to its predicted path bringing it extremely close to Earth on April 13, 2029 and again on April 13, 2036. It turns out though that while it will come close to Earth on these dates, even going inside the orbit of many of our weather satellites in 2029, it won't impact the Earth. We can even look ahead as far as April 12, 2068 with a reasonable amount of certainty, where Apophis is currently given a 1 in 149,000 chance of impacting the Earth. Ok, so we don't know of any big asteroids on a collision course at the moment, but hypothetically, if one were to hit, what actually happens?

A large asteroid has a substantial mass travelling at extremely high velocity, able to hit the Earth at speeds of tens of kilometres per second. This means it has a huge amount of kinetic energy. The largest on Earth in recorded human history was in Siberia in 1908, releasing ~10-15 megatons of TNT worth of energy, which is still quite small in the scheme of things (Shoemaker-Levy-9, which hit Jupiter in 1994, released more than 6 million megatons of TNT worth). When a large asteroid's kinetic energy is released, it is largely converted to explosive energy, heat and sound, with pressure waves travelling outwards, much like an atom bomb.

Additionally, dust, soil and rocks get thrown up into the atmosphere and out into space. For the first few hours some of these rocks would fall back down, like a storm of fireballs, their radiant energy raising the surface temperature to intolerable conditions for a short time. Finally, for species not caught in the blast and which are buffered against the cooking of the falling rocks, there's a more-long term problem to face. With enough of a dust layer in the upper atmosphere shrouding the Earth, sunlight can be blocked for several weeks or even months, acting much like a nuclear winter, preventing plants from photosynthesising and severely disrupting the food chain of the planet. Of the 5 mass extinction events in Earth's history (6 if the present-day loss of species is included), 2 are suspected to have had large asteroid impacts as contributing factors.

## Earthquakes

The Richter scale, known by many as measuring the magnitude of earthquakes, along with the decibel scale for sound, or the PH scale for acidity, is a logarithmic scale. This means that every one number higher on the scale refers to something ten times as powerful, removing the need for dealing with overly large figures and reducing the possibilities down to a small, easily managed range of numbers. Despite its well-known status though, the Richter scale hasn't really seen use since shortly after the turn of the 21st century. If you listen carefully to modern earthquake reports, you'll notice that they will say something like 'a magnitude 7 earthquake', but probably won't actually mention the scale being used. For anything above a minor or light quake (3.5 - 5.0 or so), the most likely scale they're measuring the intensity by is the moment magnitude scale (MMS). This scale is better for large earthquakes as it doesn't deal with amplitudes of the seismic waves like the Richter scale did, but rather measures the overall energy released. The reason that the replacement of the Richter scale by the MMS isn't well known is that its numbers were deliberately chosen such that at mid-range levels, they're approximately the same as those of the Richter scale, thereby easing the transition and minimising any confusion.

Misconceptions about earthquakes aren't limited to which type of scale they're being measured by though, despite being experienced by many worldwide every year, a lot of fiction around earthquakes continues to persist. For example, from various movies it's a relatively common belief that fault lines can 'open up' during an earthquake. An earthquake from a fault line is caused by friction as the ground on either side of the fault slide past each other. If the fault were to somehow open up, there would be no friction between the two sides, and hence no earthquake. Another common belief is that if an earthquake were large enough, it would be able to be a world destroyer, essentially shaking the world apart. On this front there's some good news, it can't happen. While it is technically possible to calculate the energy that such a quake would require, it's not the huge amounts of energy needed that's the limiting factor here, it's the strength of the rocks. Imagine, as an extremely simplified earthquake analogy, taking a piece of uncooked spaghetti and bending it until it snaps. Not a lot of effort/energy went into this and so the result is not a very powerful earthquake. Now imagine doing the same thing to a piece of wood the same shape. It's going to take more energy, and so this simulated earthquake will be stronger. The wood takes more energy before it breaks, and more energy released means a stronger quake. Like the spaghetti and the wood, even the strongest rocks can only take so much pressure before they break. While it's debated as to exactly how much pressure the strongest rocks could take, and hence what Earth's theoretical maximum magnitude earthquake is, all estimates generally max out at slightly under a magnitude 10, well below the world destroyer level. Just because Earth is limited in this way though, doesn't mean that everything else in the universe is. Around one in ten supernovae, instead of resulting in a typical neutron star or pulsar, results in a magnetar, a type of neutron star which spins slightly slower, taking a leisurely few seconds to complete a full rotation instead of less than one like other neutron stars, but which has a particularly powerful magnetic field, the strongest that we know of in the universe bar none, and sometimes, a magnetar can experience a 'starquake'. Earth has detected three such starquakes to date, and of these, the most powerful was in 2004, when a magnetar (named SGR 1806-20) let loose with the equivalent of a magnitude ~23 quake. So much energy was released by this that temporarily, the Earth's magnetic field was shifted, and orbiting satellites went offline. Pretty impressive, especially considering the magnetar in question is ~fifty thousand light years away.

# 19. Materials

## Night-sky cooling

You may think of ice as being a relatively modern luxury item to have on hand in warmer climates, but centuries ago, a type of building called Yakhchāls were in use in Iran to reliably produce ice in desert climates via a process known as night-sky cooling, or radiative cooling. They essentially used the same physics that can cause frost to form on the ground on a clear night. All that was required was to fill a shallow pool with water overnight, and in the morning, it would be ice which could be collected and stored. But how does water freeze overnight if the temperature never drops below freezing?

When most things cool, they emit energy via thermal radiation, as infra-red light, to somewhere cooler. We can deduce that in order to freeze, the water isn't radiating its energy to the surrounding air, so what option is left? As unlikely as it sounds, the answer is that it's radiating its energy to the cold of space. But how?

The atmosphere doesn't treat all wavelengths of light the same way. Critically, it has what's known as a transmission window. At just the right wavelengths, specifically between ~8 and 13 microns, infra-red light can pass right through the atmosphere, escaping all the way to space.

So can this strange quirk of the atmosphere be exploited beyond ice manufacturing somehow? Most definitely. Materials are starting to be developed which specifically target this transmission window. They can also be made highly reflective to the Sun's rays, so as to naturally cool during daytime as well, even when in direct sunlight. These high tech, nanophotonic materials don't have a catchy name as of yet, but chances are high that you'll see them being incorporated into existing technologies before too long, with solar cells probably being the first prime candidate. Solar cells become less efficient the warmer they get. Always something of a conundrum for something which has to be in direct sunlight to work, at least until now.

## Strength

When you think of a substance being strong, you probably think of something like steel, but the shape of something can drastically effect its physical properties. A chicken egg is designed to distribute pressure evenly all over the shell, which allows a hen to sit on it without it breaking. If you've never held an egg at the top and bottom between your thumb and forefinger and tried to squeeze it before, then you should try some time. You'd be surprised how much pressure something as fragile as an egg shell can take without breaking. The egg's shape is acting like a three-dimensional arch, with the strongest points being at the highest points of the arch (the top and bottom of the egg).

One of the hardest known substances is diamond, deriving its name from ancient Greek and actually loosely translating to unbreakable. There are different quality levels, generally though, if you want to scratch a diamond, the thing that's going to do it is another diamond, but again, shape matters. Diamonds are made of a lattice of carbon atoms, and this means they have a direction that they are vulnerable to a strike from, a cleavage plane which will cause breaking if a forceful enough strike is delivered. A skilled jeweller might exploit this weakness during gem cutting, generally though, the job of a jeweller is to ensure that if being set in a ring or the like, that the final product doesn't leave the direction that the diamond is vulnerable from exposed.

One of the most interesting examples of strength where shape matters is known as a Prince Rupert drop. This tadpole shaped piece of glass can withstand huge amounts of force to the head, you could shoot it point blank and the bullet would shatter on the glass, but the slightest amount of damage to the tail and the entire thing will explosively disintegrate. They're made by dripping molten glass into cold water. The glass on the outside of the droplet cools and solidifies immediately, while the inside of the drop remains molten for a while afterwards. As glass cools, it contracts, so as this molten core slowly cools, it evenly pulls on the outside's solid bulb shape. The outside shape is already solid and locked in, so the pulling on it from the core acts like pulling on the inside of an arch. This is essentially identical to adding pressure to the outside of an arch and so much like with the egg shell, actually makes it stronger. The inside eventually cools enough to solidify in this state with the result being the outside under extremely high compressive stresses, while the inside is in a state of extremely high tensile stress. This tensile stress forms a connected chain going all the way down the tail, leading to the explosive chain reaction should the tail be damaged in any way as this potential energy or mechanical strain energy is all released. Variations on the Prince Rupert drop technique of cooling the outside of glass before the inside is used for a variety of things in the modern world, the most common of which is probably safety glass, found in many side and rear car windows.

## Graphene

The development of a new material often comes with a lot of hype, with words like super material or the like being thrown freely around, but there is one material where the hype is probably justified, graphene. Graphene is made up purely of carbon atoms arranged in a two dimensional, hexagonal honeycomb lattice. A sheet of graphene is one atom thick, nearly transparent, conducts heat and electricity, and is also extremely strong and flexible. It can even self-repair holes in itself if exposed to molecules containing carbon like hydrocarbons. As such, it's no surprise that a range of future uses for it have been proposed, ranging from solar cells and energy storage to water filtration and even antimicrobial band aids. One of the most interesting uses for it so far though is in sound generation.

Sound generation has been using the same technology for around a century now. A microphone contains a thin piece of material which vibrates when exposed to sound waves, and these vibrations are translated to an electrical current. In a speaker, this process is reversed, with the electrical current being translated into vibrations in a piece of material which then causes the air to vibrate in the same way that was originally recorded. Until very recently, all controlled sound generation techniques used the process of mechanically pushing air back and forth to create sound waves, but by running a current through graphene, it can be rapidly heated and cooled in a very controlled way. The air in contact with it in turn expands and contracts and the result is thermoacoustics, sound waves without the need for any moving parts. While for many new materials, it's impossible to predict if they will eventually be used in future technologies or not, with graphene, it's probably more a question of how many it will eventually be incorporated into.

## Aerogels

While on the Moon, Apollo 15 astronauts demonstrated how two objects fall at the same speed when there's no air resistance, regardless of the two objects' weights. For their camera demonstration they chose to simultaneously drop a hammer and a feather. A feather was chosen as people instinctively associate them with being lightweight, with familiar sayings like 'light as a feather', but what if someone wanted to perform this experiment with the lightest material possible, what would they end up using?

In recent decades, the record for world's lightest solid has been beaten many times, but the title tends to only move from one aerogel to another. An aerogel, sometimes nicknamed 'frozen smoke' for its appearance, is an extremely porous material, first created in the 1930's by a process called supercritical drying, where the liquid components of a gel are replaced with gas. They can be made from a wide variety of chemical compounds for an array of different applications, with the title of world's lightest material moving fairly frequently due to new substances constantly being tried. However, the current lightest material (at the time of writing) has managed to hold the record for a few years running. It's an aerogel made from the previously discussed graphene.

Graphene aerogel is 7.5 times lighter than air, meaning a cubic metre of the stuff would only weigh ~160 grams (~5.6 ounces). You could balance a decent sized block of it on a flower's petals without crushing them if you chose to. Like other aerogels it's strong, flexible, and incredibly absorbent, but possibly what will end up making it particularly viable commercially is that a method has been developed allowing it to be 3D printed. It's not a super quick process, but certainly easier, faster, and more appealing than any previous aerogel creation techniques.


## Metallic hydrogen

If you ever looked at a periodic table of the elements, you may have noticed that the elements fall into groups. The right most column on the table contains all the noble gasses, left of that is an inverted staircase section containing non-metals and so on. On the far left side of the table is a column containing all the alkali metals, lithium, sodium, potassium etc., and at the top of this column, looking a little out of place, is hydrogen. But hydrogen isn't a metal, so does it really belong there?

Just because hydrogen isn't normally a metal, doesn't mean that under certain conditions it couldn't be. Under extreme pressures, it's expected that a new phase of hydrogen should be possible. The result would be a lattice of protons with predicted properties ranging from being a stable, room temperature superconductor, to being the most powerful rocket propellant option to date by far. Claims of achieving metallic hydrogen through different methods over the years have largely been met with scepticism, the most recent of these in late 2016, but this hasn't dampened the enthusiasm of high pressure physicists in the quest to reproducibly obtain a sample of this game-changing substance, and given recent advances in technology, such a breakthrough could very well happen any time now.

# 20. Matter

## Plasma

Ever huffed on a window or pane of glass and watched it fog up? You probably already know that it's just condensation from the moisture in your breath, but there's one other aspect to what's happening which is easily overlooked, the reason for the fogging up being grey is due to the moisture of your breath condensing into tiny droplets around imperfections on the glass. If you were to huff on some glass which was perfectly clean, where there were no imperfections, the moisture would have nothing to bead around, and the result would be a continuous film of moisture, which would appear to turn the glass black, not grey. No, it doesn't matter how many times you scrub at the glass with window-cleaner, you're never going to get rid of all the imperfections that way, but there is a way used in certain industries to atomically clean surfaces, the trick is to clean with plasma.

Plasma is a state of matter probably most commonly associated with the Sun, though it can be artificially created on Earth relatively easily by ionizing a gas; that is, giving a gas enough energy that you start having electrons no longer bound to their atoms. This separation of negatively charged electrons and positively charged ions (the nuclei) means that unlike other states of matter, plasma will react to magnetic fields. The higher the plasma density, or the number of these free electrons per unit volume, the stronger this reaction, with different intensities being useful for different applications, ranging from the aforementioned atomic cleaning, all the way to neon lights or toy plasma balls.

## Liquids

Look out your window, what's the first liquid you see? If you thought it was a trick question and said the window's glass, having heard that glass is some kind of supercooled liquid, you're almost certainly not alone, but you'd be mistaken. The easy guideline which you likely had to logically fight down to consider glass as a liquid is true, liquids don't hold their shape when you pick them up, glass does. The idea that glass is a liquid is actually an urban myth. It could be argued that it's somewhere between solid and liquid, a kind of amorphous solid as there is 'some' movement in its molecules as they settle over time, but the idea that it flows given enough time is false. Yes, some old European cathedrals have windows which are thicker at the bottom than at the top, but this is likely to do with production processes of the time. It's not because they've been slowly melting over the centuries, you could leave a window alone for billions of years and it still wouldn't end up in a puddle shape on the floor.

Dealing with actual liquids can definitely be interesting on occasion. For example, imagine the challenges involved when dealing with liquids in zero gravity, trying to get them to go where you want with minimal or no moving parts to help. Ferrofluids, a kind of fluid that can be controlled with magnetic fields, was one thing to come out of this area of study. More amusingly, so too was a coffee cup for astronauts aboard the international space station. Not that drinking hot liquid from open containers in zero gravity is ever going to be recommended for safety reasons, but it has been successfully done with a 'cup' that uses surface tension to direct the liquid into sippable balls near the drinker's mouth. Possibly the most interesting area of study involving liquids though concerns superfluids.

Recall from talking about non-Newtonian fluids that a fluid's viscosity is a measure of its thickness. Water is not very thick and so has quite a low viscosity, while honey is comparatively quite thick and thus has a much higher viscosity. A superfluid's defining characteristic is that it has zero viscosity, meaning that it can flow past obstacles without losing any energy through friction whatsoever. This can lead to a variety of counter-intuitive behaviours, for example you may be familiar with some liquids like petroleum, that given time will slowly creep up the side of container walls due to surface tension. Superfluid helium will also do this, but is not slowed by its viscosity. A thin film can 'flow' up and over the wall of a container with no difficulty at all. The film is so thin that capillary forces completely dominate the gravitational force which would keep a normal fluid down and in its container. Superfluid helium can also flow through tiny gaps which normal liquid helium would find too small. In theory, if you could find a way to maintain the extremely cold conditions necessary for helium superfluidity, you could stir it to create a vortex that would just keep spinning, or orchestrate a frictionless fountain that would never stop flowing. While it's still not the trickiest substance in the world to work with, a title probably reserved for anti-matter, it can certainly still make things interesting.

## Anti-matter

Anti-matter sounds like something you'd only find in science-fiction, a substance which upon contact with normal matter is completely annihilated taking with it an equal amount of matter, and leaving behind only energy in the form of gamma rays; but actually, you or someone you know has possibly encountered it before without even realising. A PET scan (short for Positron Emission Tomography) is a medical procedure sometimes used to track the progression of cancers. It works by introducing a substance called a radionuclide into the bloodstream. This radionuclide emits positrons, tiny pieces of anti-matter also known as anti-electrons. When emitted, these then annihilate a nearby electron, producing two gamma rays travelling in opposing directions. By ensuring that the substance is one favoured by cancer cells, and then detecting the gamma rays and backtracking to where they were emitted, it's possible to pinpoint where in the bloodstream the radionuclide concentrates and hence which cells are cancerous.

While currently we simply see anti-matter as a result of symmetries in the equations, our path to this rather dull sounding understanding went through quite the evolution over the years. Anti-matter was predicted years before being observed when it was noticed that equations allowed the rather peculiar feature of quantum states with negative energy. Speculating on why, if negative energy states can exist, there are no electrons which lose so much energy via emitting photons that they end up with negative energy led to the concept of the Dirac sea. The idea being that as no two electrons can share the same energy state within an atom, perhaps all the negative energy spots are already taken. An infinite sea where all the negative energy states for electrons are accounted for and there's just no room for any more. A positron would then be a 'hole' in this sea. A section where one of these infinite states wasn't already taken. This idea was left behind after the existence of positrons was confirmed in the wake of an idea sure to raise a few eyebrows, that of positrons being electrons that are travelling backwards in time. One rather out of the box idea that came from pondering this logic was that perhaps the reason that all electrons are identical is they're actually all the same entity. One electron moving forwards and backwards in time many, many times. While this idea has for the most part been left behind, when getting your head around how anti-matter behaves, the idea of matter travelling backwards in time can still be extremely useful. In Feynman diagrams for instance, used today by particle physicists everywhere to simplify interactions, a straight line with an arrow can represent either an electron travelling in the direction of the arrow, or a positron travelling in the direction against the arrow.

Anti-electrons are not the only form of anti-matter; others have been created such as anti-hydrogen, and even anti-helium. This has allowed us to compare spectral lines showing that they're identical to their counterpart elements, something which was long suspected, but incredibly difficult to prove. However, unfortunately for the science-fiction hopefuls reading, despite our relatively decent grasp of anti-matter, it's unlikely to be used as a fuel or otherwise anytime soon. Due to its current cost to produce, anti-matter is considered the most expensive substance on Earth. In 1999, NASA estimated a single gram of anti-hydrogen to cost $62.5 trillion. And then of course there's the problem of storage. You can't exactly keep something in a regular container if as soon as it touches the sides the substance is completely annihilated.

# 21. Light

## Mirages

You don't have to be struggling with thirst in a desert to see a mirage, if you've ever looked ahead along a tarmac road on a hot day, you've probably noticed it sometimes shimmers in the distance. It's the same thing. Don't worry though, it's not the heat doing things to your mind, it's the heat doing things to the light, or more specifically, the air.

While the speed of light through a vacuum is constant, through a medium it slows down. Different mediums have different speeds, different refractive indices, and when a wave, in this case light, passes between mediums with different indices, it's refracted. This is the same effect you see with a spoon partly submerged in a glass of water, the water has a higher refractive index than the air, so when looking from the side, it looks as if the edges of the spoon don't line up. Hot air and cool air have different refractive indices. On a hot day, the air just above the road is substantially hotter than the air higher up, this in turn causes the light coming towards you along the road to bend upwards. Light that was already coming from the sky at a shallow angle towards the road can bend upwards such that it misses the road entirely, not reflecting off anything before going to your eye, so what you're actually looking at when you see that water-like mirage in the distance is the sky. It may look like water, but that's only because water also reflects the sky.

143

# Reflection

Ever been in a car on a dark road and noticed how stop signs or the like really seem to shine in the headlights? Obviously anything illuminated by headlights is going to be more visible, but some signs can seem to be extra bright. This is no accident, their surface is shaped a specific way. Light travels in straight lines, and how it bounces off a reflective surface is much like a billiard ball off the side of a pool table. If you bounce a ball off an edge near a table corner, it will hit the other edge of the corner and begin travelling back roughly in the direction it came from. This is what the material of the sign is doing. If you ever look closely, you'll see the material's surface is actually made up of a multitude of tiny indentations with 90 degree corners. Instead of scattering in all directions like light would off a normal object, when the light from the car headlights enters one of these indentations, it's reflected like the billiard ball was, returning nearly all of it in roughly the direction it came in from, that is, roughly straight back at the car. This extra quantity of light being returned is what makes the sign extra bright for the car's occupants.

This way that light waves reflect off 90 degree angles was the basic reason for the initial, characteristically strange shape of stealth aircraft. If you don't have any 90 degree angles on a plane, then light waves from enemy radar, radar waves just being a lower energy, lower frequency light wave than visible light, scatter, not bouncing back to where they came from and not giving away your position. Technology continues to improve over the years, now allowing passive radar to detect if light waves are being scattered away in this fashion, but unsurprisingly stealth tech also continues to improve. Of course, one of the key ways to prevent the enemy tech from surpassing yours in this race is to not let on about all the tricks you have, and this, amusingly, is believed to be behind the story that carrots are good for your night vision. It's believed to have started as British propaganda during world war II so that they could explain away their pilot's success with night time interceptions while not revealing that it was actually due to having successfully developed and employed the use of radar (originally RAdio Detection And Ranging).

## Holograms

Take a laser, shoot it through a beam splitter, have one beam bounce off a chosen object and bounce the other off a mirror so that the two beams meet at a photographic plate, and instead of obtaining a recording of an image like you would with a photograph, you'll have a recording of an interference pattern caused by the two light paths converging. At first appearance this pattern might seem random, but by shining the same wavelength laser as was initially used through it, a 3D image of the chosen object is reconstructed, a hologram. As a hologram is a recording of not only the light's intensity (as with a photograph), but also the direction the light was coming from, with light from many different directions being recorded at every point, one of the more interesting features of a hologram is that you can cut one in half, and the entire image will still be visible through both resulting halves. This can be done as many times as desired and the entire image will still be visible in each resulting segment. You could imagine it as looking through a window at a scene. Cut the window in half and you have a smaller viewing area, but you can still see everything behind the window.

So we have the technology to view detailed 3D images using a 2D surface, but what about the kind of holograms seen in movies, the ones where it's a 3D image in mid-air? Well while we're not there yet, there are devices that allow basic shapes to be 'printed' in the air by precisely controlling the refractive index of the air in a given area. Laser light can then be refracted and reflected at different points allowing a basic 3D image to be drawn in the air. While these 3D projections technically aren't holograms as they're not recordings of light interference patterns, they may still be the next step towards the type of science fiction technology conjured in many people's imagination when they hear the word hologram.

## A random walk

How long does it take light from the Sun to reach the Earth? You may have heard it's a little over 8 minutes (the exact figure varies slightly as the Earth has an elliptical orbit around the Sun), but this is just from the edge of the Sun, or photosphere, photons of light are actually generated deep within the Sun's core. So how long does it take from when they're first created to reach the Earth?

One might assume it's a comparable time frame, but actually, the answer can be anywhere up to 50 million years (the average time estimations on this vary greatly, but the minimum suggested is still

146

around 100 thousand years), but why so long? What's the light been doing all that time?

The photon of light has been doing what physicists refer to as a random walk. Inside the Sun there's a lot of stuff to bump into. Light can't travel very far at all (measurable in millimetres) before it bumps into something and is absorbed. It gets re-emitted, but in a random direction. It then travels only a very short distance before being absorbed again. Given long enough repeating this process, through random chance a photon of light will eventually find its way to the edge of the Sun, possibly then beginning a comparatively short trip to the Earth, but due to the short distance between emission and absorption and the huge size of the Sun, the path the photon has taken during its random walk is anything but straight. It's an extremely long journey, even for something travelling at light speed.

## Pink

There was a brief but interesting discussion in the physics world at one point that pink, or more specifically magenta, was not actually a colour, or perhaps that it could more accurately be called the absence of green. For those unfamiliar with the topic, the debate went something like this.

The visible spectrum of light is made up of different wavelengths, including all the colours of the rainbow, red, orange, yellow, green, blue, indigo and violet (remembered through the acronym ROY G BIV by many a high school student the world over). When you see something in colour, you're registering either light waves corresponding to that colour, for example red, or a combination of light waves which the brain averages out to that colour, for example a combination of red and yellow registers as orange. In the middle of the visible spectrum we find green, so when you see something green, it's either due to light waves which correspond to green, or a combination of yellow and blue light waves. But what if you were to see a mix of red and violet light waves, what then? The average of these two waves is also green. The brain knows it's seeing a combination of red and violet, it has to register a single colour, but the average of these, green, is already reserved for the combination of yellow and blue, so what does it do? It makes up magenta, and that's what you see. This is why when you shine light through a prism, you get all the colours of the rainbow, but magenta isn't there.

The counter-argument to all this is that colour is not actually an inherent property of light or objects at all, but always an interpretation by the brain. In this regard, magenta is just as real as any other colour out there. They're all just made up by the brain at some level to help us interpret the world around us.

While there's more to the cases for both sides, since it starts to come down to what you mean by something being a colour, chances are this is one topic in physics that probably won't ever end with a definitive consensus.

# 22. Particle physics

## String theory

Early scholars reasoned the following: Take an object, maybe a ruler, and break it in two. Take one of the two sections and break it into two. Continue breaking these smaller and smaller sections and eventually you must reach a point where you can no longer break the object into smaller components. This hypothetical smallest component they called an atom, which roughly translates to indivisible. Clearly in modern times we've managed to look inside an atom discovering that it can indeed be further divided into protons, neutrons, electrons, and further still into quarks and gluons. But is that now the end of it, or can we look smaller still? Well, just because we haven't been able to yet, and indeed there are reasons to believe that certain things like the electron truly are indivisible, theories have been put forward speculating on what we might eventually find if and when we do manage to subdivide further, the most famous of which is a prediction from string theory.

The idea of string theory didn't just appear out of nowhere. Recall the Higgs field. How did physicists know to look for the Higgs in the first place? Why was it needed? When playing around with particles called w-bosons, something strange was observed. Everything was fine at low energies, but at very high energies, in order for the known equations to work with what was being observed, something had to be modified. These modifications in turn needed something new to work, the Higgs, and so the search was on. Now, if (and it's no small 'if' here), but if a particle, also a boson, called the graviton exists, it, much like the w-boson, would have problems arising at high energies. Can the equations be modified to fit nicely once again? The answer is yes, but when the only solution given so far is interpreted, out pops the need for the smallest things in the universe, where structures truly do become indivisible, to not be point like particles, but to resemble vibrating pieces of string.

For a short time, string theory was very much in the public spotlight when a viral video attempted to explain some of the maths behind it. Using much the same logic as assigning the value of one half to the sequence 1 -1 +1 -1 +1... etc indefinitely, it claimed that the sum of all natural numbers, that is 1 +2 +3... etc. indefinitely equals negative one twelfth. Certainly one of the strangest mathematical claims you're ever likely to encounter, and also, in the technical sense, quite wrong. In an attempt to make the maths friendlier to non-specialists, it had oversimplified something rather important. If they had just replaced the word 'equals' with something more along the lines of 'can be associated with a specific number under certain conditions, and that number is', then it's quite possible a rather large amount of negative press could have been avoided. If you'd like to understand why negative one twelfth is significant to the sum of all natural numbers without wading through complex equations, the simplest way is probably to imagine a graph, with x being all the natural numbers, and y being the sum of all the numbers up to and including x. Graphing out our sequence, we'd have part of a parabola. Now just imagine (without trying to calculate) that we extended our parabola out into -x and -y territory. Theoretically, the point where the parabola would start to bend upwards, occurs at negative one twelfth.

Badly worded maths aside, string theory can still be quite the dividing topic between physicists. There are those who see it as having huge potential, promising great things with predictions ranging from proton decay to other universes, while there are those who consider it to be more of a dead end, where every prediction has so far either been false or untestable, and expect it to contribute little or nothing of consequence in the long run. Hopefully time will settle things, but for now, the divide remains.

## The LHC

Particle physicists are nothing if not ambitious. They want to know what the universe is made of on the smallest of scales and understand how it all fits together. To date, there have been 17 unique components identified, as well as some complimentary anti-particles, that combine to make up everything in the universe. The list of 17 is comprised of 6 quarks, 6 leptons, 4 force carriers and the Higgs, and the interactions of these components is explained in detail using a theory known as the standard model. There are still mysteries such as dark matter, but for anything except gravity that you're ever likely to encounter, the standard model has it covered. But how does one actually go about finding new components to add to this list? This is where particle colliders like the LHC (Large Hadron Collider) come into play.

Particle collisions are not like any sort of collision you may be familiar with in the macroscopic world. When two high energy particles collide, all the energy involved can briefly go into the creation of a new particle if one is possible at that energy level. The new particle doesn't hang around long enough to be directly detected, it decays into more familiar particles which are what hits the detectors. As one LHC physicist[6] described it, "imagine a car collision where the two cars vanish on impact, a bicycle appears in their place, and then that bicycle explodes into two skateboards which hit our detector". The upshot of all this is that in order to potentially discover a new particle, higher, unexplored energy ranges have to be reached, which means adding energy to the particles you wish to collide in the form of speed. The particle collider capable of reaching the highest collision energies in the world by far is the LHC, capable of accelerating protons (hydrogen nuclei) to 99.999999% light speed, which is only ~11 km/h slower than the speed of light in a vacuum, before smashing them into a similar beam of protons going in the opposite direction. However, it's not as simple as just colliding two protons, saying 'we have a bump in the data, we've found a new particle' and celebrating.

Particle physicists want to be extremely sure that they're not just seeing noise, or fluctuations in the data which happen from time to time. For this, sigmas are used, the higher the sigma, the higher the degree of confidence in the results. The stringent threshold which became better known publicly after the Higgs passed it is five sigma ($5\sigma$), which means that the chance of the bump in the data being a

---

[6]James Beacham

statistical anomaly is around 1 in 3.5 million. Some use this figure (and a little mathematics) to say that physicists can then be 99.99994% certain in their claim, but that's actually a misunderstanding of what's being said. A good example is in 2011 when neutrinos were detected travelling faster than light speed with a confidence of 6 sigma. It turned out that this was due to faulty cables, but based on the data the physicists collected, 6 sigma was correct. An analogy might be trying to determine something's distance with a metre stick which is longer than a metre. No matter how careful you are or how many times you measure, the results will still be based off a faulty measuring stick. This then is one of the reasons for having multiple detectors, the more individual data sets which can confirm something, the less likely that there's some underlying problem with it. There's not just one possible result when it comes to particle collisions though, there are many, so as you might guess, reaching 5 sigma for any data set requires analysing an astounding number of collisions and to achieve this, a single detector of the LHC might register anywhere up to 600 million collisions per second, a number that certainly adds up quickly, especially when the time for a research run can be measured in months.

## Quantum chromodynamics

When it comes time to naming something in the world of physics, there have generally been two types of approach. The first is to simply name it after someone involved, for instance the Faraday cage, the Tesla coil, or the Bose-Einstein condensate. The second is to name it based on what it is, for instance a superconductor, or a microwave oven. This doesn't mean you can't have a bit of fun in the process though, and a great example of this is the gluon. Inside an atom you have protons, neutrons and electrons. Inside protons and neutrons you have things called quarks which are held together by a force. What should physicists call this force carrier that sticks quarks together, acting as a kind of nuclear glue? Why not call it a 'glue-on'? Spell it slightly differently (gluon), and *voila*, a perfectly technical sounding term to anyone that comes along, which also doubles as a bit of an inside joke.

There are six different types of quark, up, down, charm, strange, top and bottom as well as their six anti particle opposites or anti-quarks, up anti-quark, down anti-quark etc. These are designated as different quark 'flavours', and being fermions, particles with half integer spin, they follow what's known as the Pauli exclusion principle which states that 'two or more identical fermions cannot occupy the same quantum state within a quantum system simultaneously'. This can all seem a bit confusing at first, but physicists came up with a system to simplify the interactions allowing a more intuitive understanding of what's happening a lot of the time, they assign 'colour' charges to quarks, hence the area of study being called quantum *chromo*dynamics (QCD). The colours can be red, green and blue (with anti-red, anti-green and anti-blue assigned for anti-quarks) and the rules for interactions then become very easy to follow. Any combination of quarks must add up to white (from the perspective of human vision when adding colours of the light spectrum). So for instance if you have a proton (which is comprised of two up quarks and a down quark), you can assign the

154

colours red + green + blue, and the total will be white, and it's easy to see that there's no problem with the Pauli exclusion principle making it a viable particle in the real world. Likewise, if you have a meson, a particle composed of one quark and one anti-quark, you could assign the colours green and anti-green (or magenta), and again, the total will be white and there's no problems with the Pauli exclusion principle, you've got another viable particle.

Following this logic to its end, you may have noticed this means that you can't have a single quark on its own. If it can only be assigned red, green or blue, the total then won't be white and it would break the simple rule we've set out. But what happens if you rip one away from the others? Surely then you'd have a single quark right? Not so. If you imagine the quarks as marbles, then you could also imagine them as being held together by a rubber band representing the gluons, which takes more and more energy the more you want to stretch it. So if for example you took a pair of quarks, held one still and pulled the other one so far away that this rubber band snapped, then you will have put exactly the right amount of energy in to form a new quark/anti-quark pair, one attached to the one being held still, and one attached to the one you pulled out. Quarks always neighbour other quarks, it's just the way it is. As our understanding of the quantum world grows, this picture has become a little outdated, but based on it and similar analogies, in most diagrams to this day gluons are still drawn looking like springs.

The more modern understanding of what's happening is that the presence of quarks actually suppresses the normal fluctuations in the gluon field, creating areas between quarks known as flux tubes.

While quantum mechanics brings with it difficulties when talking about something's shape at this scale, for convenience you could visualise this as a straight-line tunnel dug through the gluon field between two quarks (and more of a Y shaped tunnel when between three quarks). As quarks become more separated, the flux tube between them stays the same intensity. The depth of suppression of the field remains unchanged, the energy you're putting into separating the quarks is only going into a longer flux tube. Sticking with our tunnel picture, it's like digging a longer tunnel, but leaving the diameter of the tunnel unchanged. It's these tunnels, or flux tubes then which act on the quark(s). This gives rise to quarks having what's known as asymptotic freedom. Within a particle, say a proton, they're relatively free from feeling this force, but if you try removing them from said proton, the force acting on them grows to be quite strong. Being areas where there are no fluctuations in the gluon field, you could think of flux tubes almost like vacuums (in the traditional sense of the word), attempting to suck the quarks back into the particle they're being pulled from. The bigger the tunnel dug, the more suction trying to pull the quarks back together. These tubes can't get arbitrarily long though, tunnels will eventually collapse if too long, or in the case of flux tubes, the energy required to create a new quark/anti-quark pair will be reached. Interestingly, the mass of hadrons (particles which are made up of quarks) mostly comes from the energy in these flux tubes as it takes a 'lot' of energy to clear out gluon field fluctuations. In the proton for example, only ~1% of the total mass comes from the quarks themselves.

## Quantum electrodynamics

While there may still be either high energy or weakly interacting particles left to discover, and it may be possible that we look further into particles we've already discovered finding they're made of even smaller building blocks, we can be pretty certain that the standard model lists everything that goes into making up the components of day to day life. But how can we be so sure? Isn't it possible that there's still one or more particles interacting with what we already know about but in a way we're just not comprehending? This question can be definitively answered by recalling a few things already covered in previous sections and with a quick look at Feynman diagrams.

A Feynman diagram is just a simplification of particles and their interactions, a field known as quantum electrodynamics (QED). Generally, they can represent all the involved complexities with either straight arrowed lines or squiggly lines. Remember that all particles can be characterized by their spin. Half integer spin particles, ones with non-whole number spin values are fermions (matter particles like the electron) and are designated by straight arrowed lines, while integer spin particles, ones with whole number spin values are bosons (force particles like the photon) and are designated by squiggly lines. The diagrams are then easily read from left to right and may involve something like an electron and a positron meeting, annihilating to produce a photon, which then produces another electron positron pair. Remember though that the arrow of time has no intrinsic direction to be found in the fundamental laws of physics. Much like you could record two billiard balls bouncing off each other, play it in reverse and everything would seem fine, you can also reverse a Feynman diagram and everything still works equally as well. You can even turn them 90 degrees and again, all the interactions will still be valid. This is known as crossing symmetry.

So how does this help with proving that the standard model isn't missing any basic building blocks? Well now we can draw a Feynman diagram involving our hypothetical mystery particle interacting with something we do know about, say a proton. They come in as two arrowed diagonal lines from the left, interact via some squiggly line mechanism that's not important, and scatter off each other like two billiard balls as two diagonal lines to the right. We then use crossing symmetry, rotate the diagram by 90 degrees, and we have a diagram showing a proton and an anti-proton coming together to produce our mystery particle and an anti-mystery particle a certain percentage of the time. The upshot of all this is that if there was an undiscovered basic building block particle, we could create it by smashing together combinations of the particles that we do know about, and we've looked, thoroughly. We've smashed together protons and protons, electrons and positrons, you name it and the combination's been tried, and no new mystery particles have been created. We can therefore feel secure in our original statement, the standard model covers 'all' the basic building blocks.

## Neutrinos

While the standard model might seem daunting at first, with so many different kinds of particles and interactions to keep track of, actually it's a lot simpler than you might suspect. There's really only one particle that probably needs proper introduction before the whole collection can start making complete sense, the neutrino.

Neutrinos are tiny, so tiny that for a long time it was suspected they were massless. A large quantity is produced by the Sun's fusion process, they don't have a charge, and interact extremely weakly with other things. So many are produced by the Sun and they interact so weakly that you could hold out your hand right now, and there would be anywhere up to 60 billion of them passing through just one of your fingernails every second.

Ok, so neutrinos have been introduced, but how does that help the standard model to suddenly seem simple? Well now imagine a 4 x 4 grid. In the first column, we're going to list all the easy stuff. We'll list up and down quarks, which are what combine to make up protons and neutrons, we'll list the neutrino, and we'll list the electron. All the stuff you need to make up everything in day to day life. In the second column, imagine we're listing the exact same things, just more massive versions, so in the same row as the electron in the second column we'll put the muon, which you can think of as just a big fat electron. The third column will again be exactly the same types of particles, just more massive versions, so in the third column in the same row as the electron and the muon we'll put the tau. It's similar to the electron and the muon, just more massive. Once the first 3 columns are done, we'll fill in the 4th with the appropriate force carriers, the things which don't have any mass but that carry energy like the photon or the gluon, and *voila*, a 4x4 grid that covers everything in the standard model (except the Higgs which needs its own spot outside the grid somewhere). Once you're familiar with the interactions of the first column's up and down quarks, neutrinos and electrons, the rest is pretty much just the universe copying itself a couple of times. About as straightforward as one could hope for. What's not so straightforward and still something of an unsolved mystery is determining why the universe is like that in the first place.

## Pentaquarks

Recall that quantum chromodynamics (QCD) allows a combination of quarks whose 'colours' combine to equal white. With two quarks, a meson, this could perhaps be green and anti-green (or magenta). With three quarks, a baryon, this could be red, green and blue. There's another possibility that was predicted decades ago though, something with five quarks, a pentaquark, perhaps coloured red, green, blue, green again and magenta, and their story has been something of a roller coaster ride in the world of particle physics.

A couple of years into the 21$^{st}$ century, an experiment found the first evidence of a pentaquark, named the theta plus pentaquark, which consisted of two up quarks, two down quarks and an anti-strange quark. In the following two years, eleven other experiments looked for and also found evidence of its existence. For a 16-month period beginning in early 2003, a paper on pentaquarks was published on average every second day. In 2005 however, an experiment which collected 100 times more data than the original experiment revealed that there was in fact no evidence of the theta plus pentaquark in the experiment. Despite the rigorous particle physics standard of five sigma, the bump in the data had been a random fluctuation after all, and it looked like nature for some unknown reason didn't allow the pentaquark to exist. Perhaps there was some interaction or law governing quarks that physicists weren't yet aware of. Jump forward to 2015 where analysis of particle collisions at the LHC (Large Hadron Collider) found evidence for the existence of another type of extremely short-lived pentaquark, consisting of two up quarks, a down quark, a charm quark and an anti-charm quark, dubbed a charmonium pentaquark. Many were ready for this, and again, the flood of papers on the topic of pentaquarks began, with pre-peer review theories submitted within just 30 hours of the news being released. Is that the end of the roller coaster ride though? Could it be that history will repeat itself and a more thorough future experiment will come along showing that the evidence in the data for charmonium pentaquarks was also a random fluctuation? While such a thing can never be ruled out entirely, it is extremely unlikely. Despite the LHC experiment not being designed specifically to look for pentaquarks, the data collected reached a combined significance of 15 sigma. It looks fairly certain now that pentaquarks are allowed by nature after all, and if one type can exist, it's highly probable that many others can too, we just haven't found them yet.

# 23. Gravity

## Rockets

If you spend enough time around physicists, you're likely to eventually hear one musing over a mystery, why is gravity so weak compared to the other forces of nature? There are all sorts of examples that can be given as to what they mean, for instance you can easily beat gravity briefly just by jumping, but you can't easily pull a proton out of the tip of a pencil. Another way of thinking about it might be that the strong force keeping that pencil proton attached to its neighbours is stronger than the gravitational pull acting on it from the entire Earth. However, this by no means implies that gravity is a pushover, and possibly nothing highlights this more than the difficulties involved in getting an object into space.

From the 1957 launch of Sputnik, the world's first artificial satellite, until now, rockets have been the only real option for getting an object into Earth orbit or beyond, and are quite the awe-inspiring sight to see when taking off. So much so that desks in the mission control centre face away from the windows so that people don't get distracted watching the launch they're meant to be monitoring, but if you've ever seen astronauts returning from space, you've probably noticed that it's not a whole rocket they climb out of. This comes back to the difficulties in overcoming gravity. A rocket works on Newton's third law, for every action, there is an equal and opposite reaction. Explode some rocket fuel and channel it one way, and the rocket goes the other, but once that initial burst of fuel is spent, the container for that fuel is no longer useful, it's just dead weight making it that much harder to reach the destination, so the solution? Leave it behind. You could think of the rocket more as a gun, firing its bullet upwards, but not taking the cartridge shell along for the ride. In fact the upper segment that separates does this same thing again, spending its fuel before leaving more dead weight behind. Sometimes this can happen up to four times, so the analogy becomes more of a gun, firing another smaller gun, firing an even smaller gun, which then fires a bullet, the payload. It may seem almost comical, but to date it's still the most efficient way of battling this fight against gravity that anyone's come up with. Quite recently there has been some serious interest in developing re-usable rockets, where the early stages can land to be re-used for later launches instead of just burning up and disintegrating in the atmosphere, and while there have been some very impressive successes, this technology is still in its infancy.

## Unifying forces

What would a physicist's holy grail be? What would be the one breakthrough which would top all others if it could be achieved? You probably wouldn't have to ask too many physicists before one gave you the answer 'a theory of everything'. No, they wouldn't be talking about a theory of literally everything, they'd be referring to developing a theory where all the forces in nature could be viewed as one and the same. It may initially sound a little far-fetched, but with a quick look at the history, you might find that it actually starts to sound a little inevitable.

Arguably Newton got the ball rolling on this one. Originally, the force which caused objects to fall to the ground and the force which kept the Sun, Moon and other astronomical bodies moving above seemed very different. Newton's insights attributed both effects to the one force, gravity. Later, electricity and magnetism were shown to be two sides of the same coin, electromagnetism. Similarly in more recent times, the electromagnetic force and the weak nuclear force were shown to be different aspects of the electroweak force. So what's left? How many different forces are there to unify exactly? Well, probably not as many as you might guess. There's the strong nuclear force, the electroweak force, the gravitational force, and that's it. Some theories postulate that the electroweak force and the strong nuclear force might be able to be unified at extremely high energy levels, the idea being that perhaps in the first moments after the big bang they were one and the same, but as the universe cooled, the forces split into what we see today. This would provide what's called a grand unified theory (GUT), a kind of stepping stone to a theory of everything, but the energy levels needed to test this are still well beyond the reach of current technology. Gravity on the other hand is probably going to be a 'lot' trickier. It looks like unifying it with the other forces is going to take either the introduction of something very new like an as yet unidentified field, or else some extremely out-of-the-box thinking. Despite the difficulties posed though, there are many who feel that the answer is probably out there, we just haven't found it yet.

## Space-time curvature

Thinking of gravity as a force can possibly be a little misleading. Classically speaking, a tennis ball feels an attractive force towards the Earth. When you drop it, its mass is attracted to the mass of the Earth, while the mass of the Earth is also attracted to it, but by a negligible amount in the scheme of things. Technically though, this isn't what's happening. The tennis ball and the Earth are both continuing on their original parallel paths through space, however these two paths converge. How do two parallel paths converge though, surely that makes no sense? An easy analogy would be of two ants on a basketball. If they both start off at two different points somewhere around the middle and both head 'up' the ball at the same time, they might feel that their initial paths are parallel, but due to the curvature of the ball, their paths will cross at the top of the ball. There's no force of attraction pulling the two ants together, no invisible shrinking string between them, their paths are crossing because the surface their parallel straight-line paths are traced on is curved. Likewise, the Earth and the tennis ball appear to us to have parallel paths through space, whether that's from the perspective of them orbiting the sun, spinning around the galactic centre or whatever other perspective you might choose, but much like the ants might not notice the curvature of the basketball, we don't notice the curvature of space-time. It appears to us that there's an attractive force pulling the tennis ball and the Earth together, but just like there was no invisible string between the ants, there's none here either, it's actually that the two paths are crossing due to space-time not being flat. The more massive an object, or the more energy involved, the more effect that it has on the warping of space-time around it, and so classically, the more of a gravitational force of attraction is exerted.

## The graviton

Just because we have a reasonable grasp of gravity, doesn't mean that there's no search for a deeper understanding, and to this end, theories occasionally get put forward. Over the years they've ranged from suggestions like the now mostly dismissed idea of it being a pushing force originating from all of space that gets blocked by objects to varying degrees based on their mass, to other more philosophical ones such as it being in the nature of all objects to move towards a region where time runs slower. Of all the new theories about gravity though, one stands out as having most sparked the physics community's imagination, if only briefly, the graviton.

Gravity is absent from the standard model of physics, which seems fine, the standard model works incredibly well without it, predicting the Higgs and explaining particle interactions in great detail, but what happens if gravity actually is a force in the same sense as the others? If you ever want to then incorporate it into the standard model or unify it with the other forces, you're going to have to figure out why most of its strength seems to have gone missing. One prediction along these lines is that gravity has an associated force carrier, the graviton. Gravity could then be just as strong as the other forces, but leaking into another dimension. The graviton would be a hyper-dimensional particle, so what we experience day to day in our regular three dimensions is just a tiny slice of the overall strength of gravity. This string theory friendly, game changing idea, sounds like something only a mathematician playing with abstract ideas could come up with, but it is possible to make certain predictions about such a particle should it exist, for instance its decay products would be only two photons of light, and in 2015 at the LHC, there was a bump in the data which hinted at just such a particle. While after collecting more data, it turned out that it wasn't evidence of a new particle they were seeing after all, for a brief moment in time, the graviton was by far the most discussed theory in the world of particle physics.

# 24. Magnetism

## MHD propulsion

Rockets aren't the only type of vehicle to push off something in one direction to go the other, in fact every form of transportation does this. Cars, bikes, and even just walking around pushes off the ground in the same way. Once you step off a diving board, there's no changing your mind and quickly running in the air like a cartoon character until you get back to it. With nothing to push off, it's just you and gravity.

Boats push off the water in a similar way to cars pushing off the ground, but one Japanese prototype boat from the '90s (Yamato 1) broke the mould a little in that it required no moving parts to do this. It worked using a magnetohydrodynamic (MHD) drive, which passes an electric current through the seawater in the presence of a strong magnetic field. The magnetic field then pushes the sea water out the back of the boat which in turn accelerates the boat forwards. This method of electromagnetic hydrodynamic propulsion never had a lot of commercial interest and so the maximum speeds reached were only around 15 kilometres per hour, theoretically though, the same engine principles could be applied to aircraft.

As there's no sea-water in the air to pass a current through, the trick is to ionise some of the atmosphere around the aircraft into a plasma first. This charged plasma then acts as the sea water for the MHD drive, allowing the magnetic field to push off it to achieve thrust. While predicted thrusts for such a magnetoplasmadynamic (MPD) drive are substantially better than most rockets, due to some unsolved issues, research into the area has also yet to gain commercial interest.

The idea of engines with no moving parts which use their surroundings to push off has one last surprising chapter to it, the electromagnetic drive (EmDrive). Remember those particle/anti-particle pairs constantly fluctuating in and out of existence in space? Well what if, with the right set up, you could push off them in a similar way to the MHD or MPD drives pushing off their surroundings? If possible you would then have a craft capable of thrust in space without needing to expel fuel. The amount of thrust would be tiny, but it would be continuous, substantially reducing travel times to distant targets in space. The EmDrive claims to be able to do just that. After more than a decade, testing is still inconclusive, leading to many polarised views, often split between extreme optimism and extreme scepticism, but talks of moving to a final phase of testing in space are now gaining some serious traction which will likely remove any room for speculation one way or the other. In space there's no real middle ground for the EmDrive's test results, it will either work or it won't, and time will tell us which is the case.

## Ferrofluids

Take some nanoscale particles of iron, the type you might find in the toner of some printers or photocopiers, coat them in something which stops them clumping up and sticking to each other, perhaps some citric acid, suspend them in a liquid and you'll have yourself a ferrofluid, a rather unusual type of liquid. In the presence of a magnetic field, a ferrofluid displays some distinctly non-liquid like characteristics, appearing to turn into a kind of spiky jelly, but beyond the potential for cool party tricks, ferrofluids have some extremely useful real-world applications.

Ferrofluids are less magnetic at higher temperatures. This allows them to be used for cooling without the need for extra energy input. In loudspeakers for example, a strong magnet placed near the heat producing voice coil attracts more cold ferrofluid than hot. Ferrofluids are also extremely effective at reducing friction. Coat a strong magnet with a ferrofluid, and the friction it encounters while gliding over a smooth surface or spinning on the spot will be almost negligible. Not actually zero like with superfluids, but still extremely low. It's possible that there's even a small amount of ferrofluid in your hard disk drive right now, protecting vulnerable areas by acting as a liquid seal. In fact there are so many areas where their use has major advantages that even though the study of ferrofluids, or ferrohydrodynamics, has been around for half a century or so now, there's still a fairly constant stream of new proposals being put forward for their use.

## Permanent magnets

Here's something which can get confusing quickly if careful terminology isn't used. Permanent magnets all have two ends, a north magnetic pole and a south magnetic pole. Opposites attract and likes repel, so the north magnetic pole of one permanent magnet will be attracted to another permanent magnet's south magnetic pole, while being repelled by its north magnetic pole. Straight forward enough and hopefully you're reading this section so far asking 'so where's the possibly confusing part?' In a compass, the needle is magnetised, so its north magnetic pole points north. It does so because the Earth is essentially a giant magnet, but, if the north magnetic pole of a compass points to the area of the Earth's North Pole, and a north magnetic pole of any magnet is attracted to the south magnetic pole of another magnet, then that must mean that where the compass is pointing to, the area of the Earth's North Pole, is a south magnetic pole. Likewise, the area of the Earth's South Pole is a north magnetic pole. Possibly not a good piece of trivia to bring up if people have been drinking alcohol, unless you're deliberately trying to be confusing of course.

Moving away from the pedantic and back to the physics side of things again, a permanent magnet's two poles, north and south, are a result of the electrons in the magnet being lined up, spinning in the same direction. The more that are lined up, the stronger the magnet. Cut a magnet in half, and the remaining electrons will still be spinning in the same direction they were, meaning that there will still be two poles. Every permanent magnet you've ever encountered is like this, a magnetic dipole, but that doesn't stop some physicists looking for something very different, magnetic particles with just one pole, magnetic monopoles. Certain attempts at grand unified theories (GUTs) as well as string theory would get a really solid boost should their existence ever be confirmed. As of yet there's been no reproducible evidence for them, but it takes a lot more than a lack of evidence to deter physicists from their searching.

## Destroying a magnet

There are different types of magnetism. Iron is ferromagnetic, so when in the presence of a magnetic field, its electrons will tend to line up and stay that way, itself becoming permanently magnetic in the process, but that's not the only type. Liquid oxygen for example is paramagnetic, so its electrons will align themselves in a direction parallel to an introduced magnetic field, causing it to be attracted by a magnet, but unlike with iron the liquid oxygen won't stay magnetised afterwards, its electrons will go back to their more chaotic state, not staying aligned on their own. An object can only stay ferromagnetic while under a certain temperature. This temperature is different for different materials, but if you were to heat up say one of your fridge magnets past its 'magnetic melting point', it would go from being ferromagnetic to paramagnetic. It would permanently cease to be a magnet, unless you later go to the effort of remagnetising it after it's cooled of course.

## Electromagnets

When unifying forces, how are electricity and magnetism two sides of the same coin, electromagnetism? To answer this, we'll quickly take a look at electromagnets. There's only a handful of elements which are ferromagnetic, and thus capable of being a permanent magnet, but any metal can be an electromagnet. When an electric current flows through copper wire, a metal which is not ferromagnetic, we can easily see there's a magnetic field if we have a magnet, perhaps a compass needle, close to the copper wire. When the electric current is flowing, the compass needle will be deflected. Additionally, a nearby object which is both charged and moving relative to the copper wire will experience either an attractive or repulsive force. From the charged object's perspective, this will be due to a charge imbalance in the copper wire creating an electric field. However, from an outside, stationary perspective, the copper wire doesn't have a charge imbalance, the attraction or repulsion is due to the magnetic field being produced. What's happening though, how can it be both an electric field and a magnetic field at the same time? If you think it sounds a lot like special relativity all over again, with different perspectives due to different observer's frames of reference, then you'd be absolutely right. It all comes back to the idea of Lorentz contraction.

Recall that Lorentz contraction, from special relativity, is the way that something with speed is squished in the direction of travel, so a fast-moving sphere becomes more pancake shaped from an outside observer's, non-moving reference frame. Now let's look at the piece of copper wire. Without a current passing through it, the number of negatively charged electrons exactly balances out the number of protons in the wire, so there's no charge. When an electric current is introduced, electrons flow through the wire, moving in one direction. Not at anything approaching the speed of light, but still moving, so there's still a slight amount of Lorentz contraction happening to them from an outside, stationary frame of reference. This still doesn't do anything to the charge of the wire from an outside stationary perspective though, it's still exactly balanced, just with some slightly squished electrons travelling through it. If our nearby charged object isn't moving, it won't feel a force. Finally, let's look at what's happening from the perspective of a charged object next to the wire when it is moving. Let's say the object is positively charged and moving in the direction of the electrons. For convenience, we'll imagine that it's going at the same speed as the travelling electrons. From this frame of reference, the electrons aren't moving, but the entire rest of the wire is. Not just the positively charged protons, but also the 'space' between the protons, the wire as a whole, it all gets Lorentz contracted. From this frame of reference there's now ever so slightly more protons compared to electrons in the wire. Not by much, but there are so many electrons in a piece of copper wire, and the electric force is so powerful, that this slight imbalance is enough to cause our positively charged object to feel a repulsive force from the wire. So to come back to the original question, how can electricity and magnetism be two sides of the same coin? It's because a magnetic field is just an electric field viewed from a different frame of reference.

## Lenz's law

Newton's third law, that for every action, there is an equal and opposite reaction, has something of an equivalent in magnetism, Lenz's law. It deals with the direction of current induced in a conductor by a changing magnetic field, noting that the created field will oppose the change that produced it. That might at first sound a little technical, but can be easily understood with a simple experiment. Just drop a permanent magnet through a copper or aluminium pipe. You'll immediately notice that the magnet is substantially slowed. It still falls, but at a much slower rate while within the pipe than when not. The magnet's movement as it passes through the pipe is inducing a current in the pipe which in turn produces a magnetic field opposing the falling magnet, slowing its descent.

# 25. Earth's atmosphere

## Solar flares

The Sun has something of a heartbeat, its sunspot cycle. Roughly every 11 years, due to solar material bubbling up from the core, the movement of magnetic fields inside the Sun cause its poles to flip, so north becomes south and south becomes north. After ~22 years, they return to the way they started, and the solar cycle is complete. Longer solar cycles of ~87, ~210, ~2300 and even ~6000 years have been speculated to exist, though are harder to confirm. Focusing on the shorter, clearly observed ones, this means that every ~11 years, there are a greater number of solar flares and coronal mass ejections (sometimes referred to as solar storms), powered by strong magnetic fields emerging from the Sun's interior.

The Earth is being bathed by a constant stream of charged particles released from the Sun, known as the solar wind. If this includes the highly energetic particles from a solar flare or storm, they can interact with the Earth's magnetic field causing rather stunning auroras (known as the Northern and Southern Lights, or the Aurora Borealis and Aurora Australis), but it's not all sunshine and lollipops.

Solar storms vary greatly in intensity, and technology can be easily affected by the resulting geomagnetic storms. The storm in early September of 1859, known as the Carrington event, was so strong that auroras were seen worldwide, and so bright that some mistakenly got up to start the day, assuming it was morning. Luckily, humans were not very reliant on technology then, but we did have the telegraph, and this system was knocked out, with the resulting current sparking pylons and shocking operators. Platinum contacts were nearly melted, and streams of fire from circuits started fires in offices. If one of similar strength were to hit the Earth today, transformers would burst into flames, we'd lose GPS and phone reception, power grids would be down, and we'd essentially lose in an instant access to most of our technology. Don't go panicking just yet though. While the Sun does have solar flares and storms of this strength from time to time, in fact there have even been a few recorded in just the last couple of years, the chance of the Earth being hit by one is fortunately quite slim, predicted to occur perhaps only once every 500 years or so.

## Atmospheric phenomena

Auroras are not the only natural light show to be seen on Earth. Sometimes in thunderstorms the tops of ship masts will glow in a ball of light which early sailors named Saint Elmo's fire. The 'fire' is actually plasma, caused by the presence of a pointed object in a strong magnetic field. The same effect is also believed responsible for foo fighters, balls of light that sometimes follow planes, first reported by pilots during world war II (though other theories have been put forward). In the past it was pointed out that there was some similarity, at least in looks, between foo fighters and the rather rare event of ball lightning, but appearance was shown to be where the similarities ended when in 2012, ball lightning was captured with video cameras and spectrographs. This extremely fortuitous incident allowed ball lightning to be studied in detail for the first time, revealing that the elements in the ball were the same as those found in the soil below, supporting the theory that the shockwave from a normal bolt of lightning blows vaporised pure silicon into the air which then glows while re-oxidizing.

On a clear day, with an unobstructed view of the horizon, if you watch just after sunrise or just before sunset, it's possible that you'll see a green flash, possibly lasting up to a few seconds. With the right conditions, the atmosphere acts as a prism, splitting the Sun's light and refracting the green to your eye. Due to multiple conditions needing to be met for this to occur, it's by no means guaranteed to be seen, but watching a sunrise/sunset is still probably a more appealing option than jumping a ship and heading for a thunderstorm to possibly see one of these unusual natural occurrences.

## Shadow bands

During a total solar eclipse, where the Sun's light is completely blocked by the Moon, not just mostly blocked as with an annular eclipse, for a brief period of time in the moments leading up to, and again just after the Sun is completely eclipsed, when there is just the tiniest sliver of light peeking around the side of the Moon, the most interesting thing to look at may not be up in the sky, it may be on the ground. While not visible during every total solar eclipse from every viewing area, it's possible that if you watch the ground during these moments, you'll see shadow bands, wavy lines of alternating light and dark patches, perhaps resembling ripples of light at the bottom of a swimming pool, or depending on the person describing them, possibly more like shadow snakes, moving in unison.

The most commonly accepted theory for shadow bands is that in the upper atmosphere, turbulence results in eddies in the air, which then act like lenses. The light passing through them is refracted, being split up before reaching the ground below. For distant light sources, like stars, this phenomenon can cause twinkling, but for something as big and close as the Sun, the result can instead be interference patterns, which is what the shadow bands are. There are other proposed explanations for shadow bands, some quite drastically different from this one, but even if correct, these are unlikely to be proven any time soon. Due to being a particularly rare phenomenon, only visible during some total solar eclipses, and even then only for a few moments, shadow bands are extremely difficult to study, making the process of gathering evidence to potentially support alternative theories a lot trickier than usual.


## Rainbows

Probably the easiest natural light phenomenon to observe requires nothing more than standing with your back to the Sun when it's low in the sky. Then, if it is or has just recently been raining in the distance ahead of you, there'll be a series of coloured bands forming an arc in the sky, a rainbow. You don't have to wait for conditions to be just right though, you can see the same optical effect using a garden hose spraying a mist of water, though in this case, as there's no horizon blocking the way, the arc will continue full circle allowing you to observe a ring of coloured bands.

Rainbows are just the result of light and droplets of liquid in the atmosphere. Imagine a beam of sunlight reaching a droplet. Some of the light is going to enter the droplet, continuing on until it reaches the back. Some of this light is then going to be reflected off the back of the droplet roughly the way it came, continuing on to exit the droplet roughly the same way that it entered. This is the part of the light relevant to the rainbow. When this light enters and leaves the droplet, it's refracted, with shorter wavelengths of light being refracted more than longer wavelengths of light. The result is the light gets dispersed, split into a spectrum, a continuum of wavelengths whose span includes all those visible to the human eye, though technically a rainbow is always going to be a slightly blurred spectrum of light as there will always be some 'spectral smearing'. This is due to each wavelength of light having a number of slightly different exit angles from a droplet, a result of a combination of factors, including the Sun not being a point source of light.

The exact nature of a rainbow depends on a variety of factors. For instance, different liquids have different refractive indices, so a rainbow seen in a spray of salt water, which has a higher refractive index than normal water, will have a smaller radius than a typical rainbow. Droplet size is another contributing factor. In the rare situation where a rainbow is formed at the point where two different showers with different droplet sizes have met, two different rainbows arcing out from the same apparent source point can be seen. In 2012, three different rainbows arcing out from the same apparent source point was photographed for the first time. There are a variety of other possibilities, including double rainbows, caused by a double reflection of light inside the droplets, and moonbows (or lunar rainbows), caused by moonlight instead of sunlight, often seen as white due to the faintness of moonlight compared with sunlight. Of all the possible rainbow possibilities though, it's interesting to consider that Earth may not be the only body in the solar system capable of producing them. Titan, one of Saturn's moons, is the only other place we know of apart from Earth where it rains liquid. It rains methane through an atmosphere of more than 98% nitrogen, but that's still liquid droplets falling. Add sunlight, and that's all that should be needed for a rainbow, although, receiving only around 1% of the sunlight that the Earth does, it would probably be more accurate to compare the as yet unobserved Titan rainbow to an Earth moonbow.

## Radiometric dating

The Earth is around four and a half billion years old, but how do we know this? In theory, calculating its age is a simple process. Find a rock that's survived from the time the Earth was formed (a meteorite), then measure the lead concentration in it. The ratios of naturally occurring levels of lead isotopes (Pb-206, Pb-207 and Pb-208) are well known, and the only natural process which can change these is radioactive decay, as uranium follows its radioactive decay chain to lead. Because radioactive decay happens at a constant rate, all that's then needed is a detailed knowledge of half-lives and it becomes a simple matter of calculating how long it would have taken for radioactive decay to account for any observed differences, and you've used radiometric dating to find the age of the meteorite, and hence the Earth. A simple theory, which initially turned out to be quite tricky to put into practice.

The problem turned out to be one of experiment contamination from lead in the atmosphere. This in turn was due to lead being used as an additive in gasoline at the time, a practice which had previously not been questioned. It was this method of attempting to determine the age of the Earth through radiometric dating which led to the dangers associated with the use of lead in gasoline being introduced to the public spotlight, eventually leading to the now accepted conclusion that there's no such thing as a non-toxic level of lead in the body. While today such a radiometric dating process may be comparatively easy to conduct, it was only originally feasible through the construction and use of one of the world's first clean rooms.

## Global warming

The media likes to have 'debates' about every topic. It's always possible to find someone with an outlandish belief, sit them across from anyone else in the world and then give both equal air time on a topic. Two opposing viewpoints and hey presto, must mean it's a debated topic, whether that topic be global warming or the Moon being made of Swiss cheese. This can lead some to be surprised when they discover that there's actually no debate whatsoever in the scientific community about global warming. There's plenty of political debate over it for various reasons, but absolutely no scientific debate. Humans are the driving force behind the unprecedented increase in Earth's temperature since the industrial revolution, leading to effects ranging from rising and more acidic oceans, to more extreme weather and entire species becoming extinct. But how can we be so sure it's human influence?

This can be easily answered with just a cursory look at the greenhouse effect. When the Sun's rays hit the Earth, naturally occurring greenhouse gases in its atmosphere absorb some of the energy, while the rest is radiated back into space. This is what warms the Earth. Without this natural process, the Earth would be far too cold to be habitable. More greenhouse gases in the atmosphere means more energy absorbed from the Sun's rays and thus a warmer Earth. It's then a simple step of determining what greenhouse gases humans have added to the atmosphere that wouldn't be there without our influence, and since the 1850s, due to the burning of fossil fuels which would otherwise have remained in the ground such as oil and coal, carbon dioxide in the atmosphere has risen from 280 ppm (parts per million) to now over 400 ppm. In the long term, one of our closest planetary neighbours, Venus, shows what happens if such a trend is left unchecked, as it's a planet extremely similar to Earth in both composition and orbit, with the only difference being that it's in the grips of a runaway greenhouse effect. There may be a lot of speculation when it comes to the more short to mid-range predictions of future effects due to global warming, but when it comes to the reason behind the scenes, there's no room for speculation, it's humans.

# 26. Number crunching

## Entropy

Energy arrives at the Earth via photons from the Sun, and give or take a bit of global warming, we radiate back exactly the same amount of energy into the universe. However, for every 1 photon we take in, we radiate back 20 photons, each with roughly 1/20$^{th}$ the energy. The energy of the system remains the same, but the entropy of the system is always increasing as time passes. This is just one example of how entropy applies to any system you care to name. It's more than just physicist book keeping, it's at the heart of ideas ranging from negative temperatures to the arrow of time. Because there are always more ways for a system to be disordered or have high entropy than there are for it to be ordered or have low entropy, a system will always tend towards higher entropy given time. However, what happens if you look at the complexity of a system over time?

This question doesn't have to come purely from the perspective of scientific curiosity, it can also come from a more philosophical base. You might for instance ask that if it's simply in the nature of any system, for instance the universe, to go towards being more disorderly, why are we here? How did anything complex like humans ever emerge? The answer is that while entropy is predictable, going up over time, complexity is also predictable, but as time goes on it first goes up and then goes down again. For an example of this, imagine the standard cup of coffee with cream on top. This is a very ordered system with low entropy. It is also quite simple with quite a low level of complexity. Once the cream is fully mixed into the coffee, it will have a high entropy, but again this arrangement of its atoms is still quite simple with a low level of complexity. Now imagine a point of time in the middle of these two extremes, where some tendrils of cream have started to mix into the coffee, but it's not yet fully mixed. At this point in time, the entropy is somewhere in the middle, but the system is quite complex. If you wanted to model the position of the various atoms, it would take a lot more work than if it was just cream on the top, coffee on the bottom, or if it was even amounts of cream and coffee throughout. As time goes on, a system always goes from being simple to complex and back to simple again, and the universe is no exception. Entropy is taking the universe from the simple ordered state at the moment of the big bang to the simple disordered state of thermal equilibrium during the dark era, but in the meantime we find the stelliferous era which is where the universe is currently at, a time when the complexity of the system is extremely high.

## Natural units

You can state the speed of light many different ways. Perhaps you have a head for details and can remember 299,792,458 metres per second. Perhaps you prefer to round it off to $3 \times 10^8$ meters per second, or perhaps something that sounds a little catchier appeals like 1 foot per nanosecond. But even when using just the letter c (for constant), physicists can still easily find themselves writing it, as well as other things like ħ (the reduced Plank constant, equal to $6.58 \times 10^{-16}$ eVs) in seemingly endless combinations in their equations. To get around this, they sometimes opt to use what's known as natural units. Unlike any other system, natural units are based purely on universal physical constants. With a little playing, things can then be made a lot simpler by adjusting it such that $C = ħ = 1$ before beginning any calculations, only ever returning to regular units at the end once an answer has been obtained. Other constants besides $C$ and ħ can be adjusted, though it's not always a viable option.

To quote Planck, "we get the possibility to establish units for length, mass, time and temperature which, being independent of specific bodies or substances, retain their meaning for all times and all cultures, even non-terrestrial and non-human ones and could therefore serve as natural units of measurements". So there you have it. Not that humans will necessarily ever find themselves talking to aliens about the time of day or how hot it is, but if we do, we already know the unit system we should probably use.

## Big K

While natural units are extremely useful in calculations, it is still the metric system, working on the International System of Units (SI), that's used to allow one scientist's units of measurement to be as close as possible to any others. As technology has progressed, so too has our ability to be more exact in said measurements. For example, the scientific definition of a metre is currently the distance light travels in a vacuum given 1/299,792,458th of a second, a rather precise definition. Practically, if you wanted to measure an exact metre, you could take a helium-neon laser, fire it in a vacuum and count out exactly 1,579,800.298728 wavelengths. This level of accuracy and reproducibility means that no matter where you are, if required, it's possible to determine an extremely precise metre given the right equipment. Until 2019 though, there was a long standing problem with the metric system, its lingering technical reliance on the International Prototype of the Kilogram (IPK), sometimes referred to as Big K.

The IPK is a 19th century, one kilogram weight made of a platinum/iridium alloy, kept in a vault in France. So what's the big deal? What makes it so special compared to any other one kilogram weight in the world? The reason is that technically, until the change in 2019, all other weights in the world were measured against the IPK. It's not that the system had specifications for working out one kilogram like with working out one metre and that's what the IPK was made to, it's the opposite. One kilogram was actually defined as being equal to the weight of the IPK. A lot of work went into moving on from this rather archaic relic from the past, with a main stumbling block for a long time being the difficulty in finding a way to realise one kilogram in the lab with a high enough level of accuracy. Multiple proposals of atom counting, along with a few others were suggested, but in the end, the accepted solution was to use the ratio of energy to the frequency of the photon. That may sound complicated, but all that's really needed is a knowledge of the Plank constant. The Planck constant can be calculated via three measurements, length, mass, and time. If one sets the numerical value of the Planck constant at 6.62607015x10^-34 Joules, then combined with the already accurately defined metre and second, the kilogram can then be calculated. The IPK will continue to be stored in its vault, an interesting relic of the past, but no longer the only exact measurement of the kilogram available to the world.

## The parsec

Metres are fine for measuring distances, but when it comes to distances in space, the numbers can quickly become so large as to make them impractical, a new unit is needed. The average distance (due to Earth's slightly elliptical orbit) between the Earth and the Sun is known as an astronomical unit (au), a convenient unit translating to just under 150 million kilometres, but what about for the really big distances? While it would be possible to use something like light-years, the accepted convention for various reasons is to use the parsec for great distances. It was previously mentioned that the radius of the observable universe is ~28.5 gigaparsecs, a gigaparsec being 1 billion parsecs, but what exactly is a parsec?

The word parsec comes from an abbreviation of the *par*allax of one arc*sec*ond. If you were to imagine a right-angle triangle drawn between the Earth, the Sun, and a distant object in space, with the hypotenuse of the triangle being the line from Earth to the distant object, the further that object is away, the smaller the distant angle of the triangle becomes. When that distant angle is 1 degrees, the distance to the object is 1 parsec. For those with a head for maths, this is exactly 648,000/π astronomical units, or for those wanting it in more traditional units, a little over 19 trillion miles.

## Calories

People can be surprised to discover that the calorie (cal) is an obsolete unit of measurement, not generally used in the sciences since the introduction of the International System of Units (SI) mid last century. The accepted unit for measuring energy is now the joule (J), but what was wrong with the calorie? Why didn't it make the cut?

A calorie is the amount of energy required to warm one gram of water one degree at one atmosphere of pressure. Sounds simple enough, but just from this description there are uncertainties creeping into the works. The amount of energy required will vary slightly based on the starting temperature of the water, and even given a starting point, achieving 'exactly' one atmosphere is no easy task, with variations again having effects on the outcome.

There is one more amusing detail which is certainly not in the calorie's favour, the fact that it can be acceptable to call 1000 calories a Calorie (Cal). Sometimes there's a distinction made between the two, one being the small calorie and one being the large calorie, but sometimes the only distinction being made between the two is the capital C, so 1000 calories = 1 Calorie. Today, while there are some chemists who might still occasionally use the calorie if it's more convenient, the only real use of it as a unit of measurement is in the food industry, most notably in the U.S.

There's no particular reason for this, it's just what they've decided to go with, much like being one of three countries in the world to not yet officially adopt the metric system. The small amount of wriggle room in the definition may even be seen as an advantage by some for their products. While there is certainly some confusion around the Calorie, sometimes even reaching the point where it is confused with the kilojoule (a kilojoule is actually equal to ~200 cal or ~0.2 Cal), if a food is listing its nutritional content and using calories, since almost all foods nowadays like to be seen as having less, it's generally safe to assume that they're talking large calories (Cal), whether or not they remember the capital C.

# 27. A touch of philosophy

## Planck units

There are many different types of logical proof. Probably the first examples of the idea of 'proof by contradiction' were a series of paradoxes generally credited to Zeno in the 5th century BC. There were multiple of these focusing on similar ideas, but arguably the most famous involves a race between Achilles and a tortoise. It assumes the tortoise to have a head start, and then points to the problem that for Achilles to overtake the tortoise, he must first reach the point where the tortoise is now. Once he reaches this point though, the tortoise has moved on, perhaps a tenth that distance again, and so he must again reach where the tortoise is before being able to overtake it. While the distance to the tortoise is always getting smaller, this repeats indefinitely as further divisions are always possible, and so Achilles can never overtake the tortoise as it would always involve an infinite number of steps, consisting of continually trying to reach the point ahead where the tortoise is currently. Clearly the conclusion is easy to disprove, something can be overtaken by something else travelling faster than it, but that's missing the point. In order to fully solve a paradox, one must show what is wrong with the argument. There were many solutions proposed over the centuries, but it wasn't until the early 20th century when physics hit upon the idea of Planck units that many felt comfortable with calling the paradox completely solved.

The idea of Planck units is that there is a smallest measurable unit for things, below which measurements have no real meaning. Relevant to the Achilles and tortoise paradox is the Planck length, the smallest measurable length, indivisible into further segments. This takes away the infinite number of steps Achilles needs to finally reach the tortoise as there is a point where the distance between them cannot be divided further. Mathematically, this length is $1.616229 \times 10^{-35}$ metres, but if you want to be able to visualise it, imagine a single strand of hair, expanded out to the size of the observable universe. In this extremely stretched out space, a Planck length would then coincidentally be around the width of a single strand of hair. Aside from length, Planck units also cover mass, time, charge and temperature. The idea of these units may seem strange at first, but they start to make a lot more sense once you look at where they come from. At their core they're just the effects of living in a universe where light is quantized (is made up of individual photons). Take the idea of Planck temperature for example. An individual photon of light carries with it a set amount of energy. If a particle loses heat, it loses it by emitting a photon of light. So once a particle's energy is so low that it can no longer emit a photon of light, you've found the Planck temperature, and talking about dividing its energy (reducing its temperature) further in this way loses any real meaning.

While the concept of Planck units is fully accepted and understood, they do bring with them their own questions. One which is heavily debated is whether or not the Planck length can be applied to space. Is space a continuous backdrop, or is it grainy, made up of a multitude of tiny, indivisible segments? Currently we have no way to answer this. Ironically, Planck units have also lead to modern paradoxes much like Zeno's being put forward. For example, imagine a right-angle triangle at the Planck scale. Imagine that the hypotenuse is exactly one Planck length long. Now ask yourself, how long are the triangle's sides? They can't be smaller than a Planck length, but the Pythagorean theorem is pretty clear that they also can't be equal to or larger than one.

## The anthropic principle

Ever been in a conversation with a curious child where no matter what answer you give the next question they ask is why? Of course this is fine for a while, but the inevitable problem soon shows itself, that the question of why can always be asked one more time than you have answers. This is true for anything, and physics is no exception. If you keep asking why, you'll eventually reach a point where answers are still being searched for, and it's in this realm that you can occasionally find some ideas which are more philosophy than physics popping up. It may cause many a physicist to groan just at the mention of it, but the most common of these ideas to encounter is probably the anthropic principle.

There are a few different versions of the anthropic principle, but the basic idea always remains the same. When asking something like why the cosmological constant is exactly the value that it is, where if it were just a little different then life as we know it probably wouldn't exist; or perhaps something simpler like why is the age of the universe what it is, where of the infinite stretch of time that it will be around for we seem to find ourselves in the incredibly slim portion which is hospitable to stars and life, the anthropic principle says that the reason for all these things is that if the universe wasn't like this, that conscious life wouldn't be able to observe it. Sometimes it goes further to suggest that much like in string theory, there may be a multiverse of possibilities, many universes all with different cosmological constants and the like, and so it's only in a universe where everything is just right that these things can ever be observed. The answer to any question of why then essentially becomes because we can observe it, it must be compatible with intelligent life. At this stage, you can probably see why physicists might groan at the mention of the anthropic principle. Not only is it essentially untestable as well as easily seen as a tautology, essentially just saying that 'if things were different, they would be different', but it also tries to give a final answer. Something which definitely goes against the idea of physics being the never-ending search for more answers that many see it as.

## Ridiculous extremes

Occasionally in physics you'll come across a thought experiment which seems downright ludicrous on the surface, but actually, these can be purposefully designed to be that way in an attempt to highlight a perceived problem with a particular theory. The most famous example of this would probably be Schrödinger's cat, a thought experiment where a cat is placed in a sealed box with a vial of poison linked to a radioactive source which may or may not decay at any given moment triggering the vial of poison to break. This

194

gedanken, or thought experiment deals with superposition, by manufacturing a state where at any given time before looking in the box, the cat would be in a superposition of states, being both alive and dead at the same time. Of course it was never proposed as an actual experiment, merely put forward as an argument against quantum mechanics in the early days of the field's development. There are now many ways of looking at what would actually be happening, but the simplest you've probably already thought of. Surely the cat would know if it's still alive.

Another, possibly less well-known example deals with Boltzmann brains. The idea stems from pondering why the universe was in such a well-ordered state in the past. One theory put forward very early on was that given an infinitely long time span, random fluctuations could account for it, so we could just be living in the aftermath of an uncommonly large random fluctuation. The idea of Boltzmann brains played on this by pointing out that compared to all the matter in the universe coming together into a well ordered, small area via an extremely large random fluctuation, it was overwhelmingly more likely for small fluctuations to occur; taken to extremes, perhaps that enough matter would come together to make up just the room that you're in right now, or indeed, just your brain, along with all your memories and thoughts of having a body, a room around you and a world outside. Unlike the Schrödinger's cat example, this succeeded in disproving the random large fluctuation theory. While it's true that at any point you can ask 'am I a Boltzmann brain?', and to answer it any previous determinations to the contrary must be put aside as they may stem from memories which have just formed; as a physics professor[7] once said, "you don't walk out tall buildings from the top floor, likewise to live in the world, you have to make some sense of what you see around you, and to do that, the only intellectually respectable thing is to imagine that you are not a Boltzmann brain".

Sometimes these thought experiments can lead to testable predictions as well, whether or not that's the intention of the person behind them. If you have a point-like source of light illuminating a circular object to cast a shadow, and you have just the right ratio of diameter of the object, the wavelength of the light, and the distance between the object and the screen for the shadow, in the middle of the circular shadow there will be a small point of light. This is caused by light diffracting around the edge of the object. When this happens, every point around the outside of the object can then be considered a new point of light, all with exactly the same distance to the centre of the shadow where they constructively interfere with themselves to make the bright spot. This is often referred to as Poisson's spot, probably much to the dismay of its namesake Siméon Poisson if he were still alive today. While an excellent prediction, he originally pointed out that it would be a consequence of light travelling as a wave in order to ridicule and dismiss an entry for an 1818 competition to explain the properties of light to which he was a judge. Poisson's prediction was tested by another judge and shown to be accurate, convincing scientists of the day of the wave-

[7]Sean Carroll

like properties of light, with the name Poisson's spot sticking ever since, an unfortunately lasting reminder of a moment of hubris.

## Tesseracts

It's easy to say something like 'space can be curved', followed by talking about the effects that this can have, but actually coming to grips with the core of the idea, of a dimension beyond our realm of experience, can be a little trickier. For many though, once comfortable with the basics, this concept can prove to be endlessly entertaining as there's no real limit to the level of complexity which extra dimensional thought experiments can be taken to. While it doesn't sound like it initially, the simplest starting point for all this is probably to consider a tesseract, the name given to a four-dimensional hypercube.

To properly get a feel for what a tesseract is, it's easiest to work up through the dimensions. Take a line segment, and move it at right angles to itself an equal distance. You now have a square. If you take that square, and move it at right angles to itself an equal distance, you have a cube. Now before jumping ahead and trying to visualise what happens when you take a cube and move it at right angles to itself an equal distance, first look at the shadow of the cube in two dimensions. Assuming a mostly transparent cube, the shadow that it casts will be of two squares, connected by four lines from their corners. This picture of a cube is simple to draw, but if you examine it closely, you'll notice that not all the lines are equal, and not all the angles are right angles. This is part of the cost of representing a three-dimensional object in two dimensions, it isn't perfect, a shadow loses things by projecting something one dimension down. Now we're ready to move the cube at right angles to itself an equal distance, resulting in a tesseract. While this can't actually be done, humans being three dimensional creatures and all, we can still visualise what the tesseract's three-dimensional projection would look like, it's three-dimensional 'shadow'. Imagine a small cube, nested inside a larger cube, with the eight corners of the smaller, inside cube, connected via edges to the eight corners of the larger, outside cube. Unlike the four-dimensional tesseract which would have all right angles and edges of equal length, this three-dimensional object has some edges which aren't equal, as well as some angles which aren't right angles. In the same way that the three-dimensional cube lost things when being represented in two dimensions by its shadow, we've lost things representing the four-dimensional tesseract in three dimensions. It's just the cost that needs to be paid when projecting something one dimension down though.

# 28. Thought experiments in space

## Kugelblitzes

Light has energy, energy has gravity, enough gravity and you get a black hole. So the question then arises, is it theoretically possible to have a black hole made entirely of light? While there's no known natural process that could come close to causing this, there's still a name for such a thing in theoretical physics, a kugelblitz, not to be confused with the second world war anti-aircraft gun the Flakpanzer IV Kugelblitz. While it's a fun idea to consider, it's unlikely that the kugelblitz will ever move from the realm of the theoretical to that of observed reality. Not only would a kugelblitz be indistinguishable from a regular black hole once formed, but making one would take a 'lot' of light. It would take so much in such a small area that the heat involved during its formation would actually be above what physics is able to cope with. More accurately, the region would be so hot that electromagnetic waves radiating out from the region would have a wavelength shorter than the Planck scale. A definite problem, but one which would not be totally unique in the world of physics. It's also faced by those trying to grapple with what was happening in the universe during what's known as the Planck Era, the time before the first $\sim 1 \times 10^{-43}$rd of a second after the big bang. Our understanding of the universe is great after this point, but before then, the laws of physics as we know them tend to stop being applicable in the way that we'd like.

## The Drake equation[8]

Suppose you wanted to estimate the chances of other advanced civilizations currently living in the Milky Way, at least advanced enough for radio astronomy, how would you go about it? While there are obviously a lot of unknowns, so the answer isn't going to be exact by any means, it can still be thought about in the following way.

Being conservative where possible, but with numbers also chosen for their ease of calculation, we first take the number of stars in the Milky Way ($N_*$), estimated somewhere between 100 and 400 billion, let's then choose the figure 200 billion. Next is the fraction of those with planets ($F_p$), and while most stars are suspected to have planets, let's be extra conservative here and call this 1/2, so 200 billion x 1/2 = 100 billion. Next is the number of planets per system which might have habitats ecologically suitable for life ($N_e$). This one is trickier. Our system for instance has Earth obviously, but there's also Mars, Titan (one of Saturn's moons), the list goes on. Our system might not be indicative of what's out there though, so to again be conservative we'll say 2, so 2 x 100 billion = 200 billion. Next is another tricky one, the fraction of times life arises on such a planet ($F_l$). We currently only have one planet to go from, the Earth, where life arose extremely quickly once given the possibility, and until we thoroughly explore more of the solar system, this is all we can go by. As there may be impediments to this explosion of life happening elsewhere which we don't know about though, and we're

---

[8]This section was greatly influenced by an explanation given by Carl Sagan.

trying to stay conservative, we'll say for this 1/2, so 1/2 x 200 billion = 100 billion. Next, we ask what's the fraction of these planets where intelligence develops ($F_i$), and for the chance of such intelligence developing communicative technology ($F_c$), at least to the level of radio astronomy. While a staggering number of things needed to happen on Earth for human intelligence to develop such technology, there may be many paths towards the same end. These two then have estimates all over the place, but taking some middle ground for the estimates, we'll go with 1/10 for both cases, so 1/10 x 1/10 x 100 billion = 1 billion. Finally, for what percentage of a lifetime of a planet is such a civilization present (L)? While we don't know how long humans will be around for, as we could easily wipe ourselves out in the not too distant future, we can say that the relevant technology has only been on Earth for a matter of decades, while the lifetime of the planet has been a few billion years. If this is typical, then we can go with the number 1/100 million. 1/100 million x 1 billion = 10, so our final conservative answer to our equation of ($N_*$) x ($F_p$) x ($N_e$) x ($F_l$) x ($F_i$) x ($F_c$) x (L) = ~10. ~10 civilizations, including us, currently in the Milky Way capable of communicating. Not a very large estimate considering the size of the area.

This equation is known as the Drake equation. Sometimes you'll see different symbols used, but they're just for convenience, they're not locked in stone. Sometimes the equation is played around with, removing different variables so that one can talk about the number of times life might have evolved in the Milky Way, or the number of planets with not just intelligent life, but life of any kind that might presently be out there. Due to the huge influence of unknown variables, estimates produced from the equation vary wildly and aren't taken super seriously, but the equation is still a useful tool as it addresses all the relevant factors when pondering the possibility of extra-terrestrials, while also highlighting which areas still need to be addressed if we're ever to be able to give a solid estimate of our chances of detecting the signals of such beings.

## The Oort cloud

There are a lot of ways for comets to die. They frequently cross the paths of planets and collide with them. Alternatively they can be gravitationally ejected from the solar system into outer space. And even if neither of these two things happen, they're mainly made of ice, so after enough trips close to the Sun, this will all evaporate away leaving behind just an asteroid. Despite this, the number of comets we see doesn't seem to be diminishing. Analysing their orbits they don't appear to have come from interstellar space, and they can come from all directions, there's no preferred angle, so the question is raised, where are they coming from?

While never directly observed, the accepted hypothesis is that around the solar system at a distance of between 5,000 au and 100,000 au is something called the Oort cloud, a vast collection of icy objects, formed early in the solar system's history, but flung out in all directions by gravitational interactions with the planets and gas giants. Due to this method of formation, the Oort cloud doesn't follow the normal rules. It's not a flat disc shape, but spherical. Guesses as to the exact nature of the Oort cloud can vary based on a few different factors, but estimates between one and two trillion objects much like Halley's Comet are not uncommon, giving the Oort cloud a combined mass around five times that of Earth. There are some alternative proposals for the way in which the Oort cloud formed, but no alternative proposals for where all the long-period comets come from. Despite the idea now being over half a century old, the Oort cloud is still our best guess. Unfortunately we'll never see the Oort cloud as a whole as it's dark in space, and each object is expected to be about as far from its closest neighbour as Earth is from Saturn, but when one of these objects makes the long trek into the inner solar system, it can make up for this with quite a show. Comets can sometimes be seen from Earth without the aid of a telescope, with tails that can stretch for an entire au. An impressive sight, interpreted in a variety of ways by different cultures throughout the ages.

# 29. Dabbling in space

## Voyager

The most well-known of all man-made probes are likely the twin Voyager probes. Launched in 1977 to take advantage of an alignment of the outer planets of the solar system, an alignment which occurs once every 175 years, they used this to perform a series of gravity assist manoeuvres, sling-shotting from body to body in the solar system. In 1990, Voyager 1 turned around to take the first ever picture of the solar system from such an outside perspective. It included Earth, the 'pale blue dot', as seen from ~40.5 au (~6 billion kilometres or ~3.7 billion miles), before cameras were shut down to conserve power and both craft continued on with their interstellar mission. This part of their travels took them out towards the heliosphere, the bubble-like region of space which protects the solar system from a large amount of cosmic rays. It's created by the Sun's solar winds pushing out against the interstellar medium. The exact shape of the heliosphere is not static, it's constantly changing over time, perhaps growing due to solar flares, or shrinking due to being pushed back by the effects of a supernova. In 2012, Voyager 1 crossed the outer edge of the heliosphere, known as the heliopause, becoming the first craft to ever enter the interstellar medium (ISM), or interstellar space. Many people know of the 'golden record' carried by the Voyager craft, the gold plated audio-visual disc containing a range of things including greetings from Earth in various languages, but while any advanced civilization listening to the record in the far distant future may not understand what's being said on the record, chances are high that they'll be able to operate it, as well as be able to figure out where it came from due to the cleverly designed cover of the record. But how is it possible to convey such detailed information without a common language?

The key is to first establish a unit of time which is universally understood. For this, a simple picture of a hydrogen atom, the most common element in the universe is depicted. It's depicted twice, showing its electron in two different spin states. The electron in a hydrogen atom flips from spin up to spin down and back again at a constant rate, so through this one small diagram, assuming a base level of science on the part of the reader, a unit of time has been established. For the rest of the diagrams, a binary code is used, with a 'I' being a binary one and a '_' being a binary zero. So around an overhead picture of the record is a series of Is and _s which indicate to the reader how many time units there should be per revolution of the record. Underneath a side-on picture of the record, a measurement of how long a single side of the record should take to play. Other instructions for the record's use indicate how to get visual information out of it by showing a waveform with a time measurement, a video image frame with vertical scanning to show the direction and time of each scan sweep, and the same frame again, but with a circle in the middle, the first image that they'll get should everything be done correctly. The choice of a circle also allows them to calibrate settings to get the correct ratios for the rest of the disc's images. This just leaves the final image on the cover, the one telling the reader the probe's origin, our galactic address, and this is possibly the cleverest of the lot. It's a diagram which defines the position of our Sun according to the position of the closest fourteen pulsars. Different distances to pulsars are represented by different length lines radiating out from a central point, with each line containing the information, in the established binary time measurement, for the frequency of the associated pulses, the regular time interval between bursts of electromagnet radiation which is unique to each pulsar, being based on their exact rotational speed.

While the probes have long journeys ahead of them, not even reaching the Oort cloud for another ~300 years, and not passing through it completely for ~30,000 years, and while the chances are against them ever being found by intelligent beings at all, it doesn't really matter one way or another where they end up, the fact that humanity managed to get them there at all is an achievement in its own right. Carl Sagan possibly summed this up best, saying "the launching of this bottle into the cosmic ocean says something very hopeful about life on this planet".

## Cold welding

Normally if you want to weld two pieces of metal together, you'll heat where you want to connect them until you have a liquid, possibly add a filler material to the molten area, press the two pieces together, and when the molten metal cools, you have a joint. There is another way possible though. If you have two pieces of similar metal without anything covering the ends, as in no oxidation layer, which all metals exposed to air are covered in, no grease from fingerprints, or dirt, or water vapour etc., just the exposed metal, then when you bring the two pieces together, they have no way of knowing that they're two distinct pieces of metal. They become one piece of metal via a process known as cold welding, though generally the 'weld' isn't as strong as a regular weld. This is because while, where the two pieces touch, the connection is better than a normal weld, being essentially indistinguishable from one continuous piece of metal, the two pieces typically aren't atomically smooth when brought together. Microscopic rough patches prevent them touching all the way along the join, and extremely close to touching isn't good enough for cold welding.

While cold welding is incredibly difficult to achieve on Earth, even with an ultra-high vacuum (UHV) system, when launching something into space where it's naturally a vacuum, it is something which has to be accounted for as it can easily happen unintentionally. If you have a moving part which needs to stay moving for example, keeping it well lubricated is extra important. Something which humans had to learn the hard way, with problems persisting through the decades from early space missions in the '60s, to satellites like Galileo, launched in 1989 to study things including Jupiter and its moons, but which failed to properly deploy its antenna due to suspected cold welding from a lack of lubricant.

## The smell of space

There's a lot of things astronauts have to deal with when in space for long enough, like when living on board the ISS (International Space Station). There are the temporary physical changes due to a lack of gravity, like their spines stretching allowing them to grow a few percent taller, the body losing muscle mass and bone density, or liquids in the body redistributing differently resulting in a puffy looking head. There are the lifestyle changes, like getting used to a wet towel instead of a shower, a urine processor for recycling water, or a sunrise happening every 90 minutes. There are potentially long-term problems from exposure to high energy cosmic rays which currently can't be protected against and are responsible for the 'flashes' of light that astronauts report seeing, thought to occur when the rays interact with the eyes and/or brain. Then there are the little quirky things that are really just part of the experience, like the odd sensation of waking up in the middle of the night and not knowing which way is up, or discovering that space has a distinct odour.

That's not to say that the space station itself smells, though it's perfectly possible that it does. The Mir space station was once described as smelling like a mix of sweaty feet, stale body odour, and thanks to the cosmonaut love of vodka, nail polish remover and gasoline. No, when astronauts return from a spacewalk and remove their helmets, their suits and tools have a distinct smell clinging to them that wasn't there before. While a couple of explanations have been put forward, the most likely culprit is thought to be PAHs (Polycyclic Aromatic Hydrocarbons), originating from the Earth's atmosphere. PAH's are produced from the incomplete combustion of organic matter, and so the Earth has an abundance of them, from tobacco smoke and car fumes, to incinerators and forest fires. These particles when brought back inside the station then mix with the air resulting in the distinctive odour. While it's been described a multitude of different ways, some of the most common themes across all the descriptions of the smell include 'burnt', 'metallic' and 'meaty'.

# 30. End on a joke

A useful tool for physicists is the ability to simplify something down so that they can focus on a particular aspect of a problem. It would be no surprise on a randomly chosen physics exam to find a question asking you to figure out what happens in such and such a scenario, ignoring friction, a degree of simplification that other professions might balk at.

This level of simplification when tackling problems regularly gets made fun of with various jokes, but amusingly, many come back to the same punch line. In fact, just the last five words of the punch line can be used as a way of light heartedly suggesting that someone is oversimplifying a problem they're working on. Generally the full joke involves a scenario where a physicist is called in to help or analyse a problem, perhaps a farm where the chickens have ceased laying eggs. After whatever amount of build up the specific version of the joke entails, it ends with the physicist exclaiming...

"I have the solution! But it requires spherical chickens in a vacuum".

## Acknowledgements

Cover photo: NASA's Hubble space telescope image of NGC 1672, a barred spiral galaxy located in the constellation Dorado.

While this has been a book of (hopefully interesting) facts, the nature of which means that most do not stem from any one particular source (which papers should one use to reference the speed of light in a vacuum?), the way certain facts have been summarised, as well as some of the analogies used when doing so sometimes can be traced to specific sources of inspiration. As the nature of summarising lends itself not to papers but talking, the three largest such influences by far have been internet videos, talks and lectures, and while links come and go online rendering any list of specific web addresses relatively useless in the long run, I would still like to acknowledge, as well as to recommend for any readers who found themselves liking this book and wanting to search out more such facts on their own the following.

For lectures, the physics for future presidents (PFFP) course from Berkeley is a must watch. Some of the free MIT online courses are also extremely good.

For talks, there are many good science related TED talks out there. Occasionally there's also a science-related Google talk, not as often, but the ones that they do have tend to be extremely good.

Lastly for videos, while there are many good options, with several providing a multitude of extremely short (~5 minute) videos succinctly summarising a variety of interesting things from the world of physics, my two personal favourites are 'when the apple drops' and 'minute physics'.

www.ingramcontent.com/pod-product-compliance
Lightning Source LLC
Chambersburg PA
CBHW030620220526
45463CB00004B/1354